COAL—
Bridge to the Future

Report of the World Coal Study

Carroll L. Wilson
Project Director, Massachusetts Institute of Technology

Ballinger Publishing Company • **Cambridge, Massachusetts**
A subsidiary of Harper & Row, Publishers, Inc.

Cover with title in the 13 languages of WOCOL members on background color of burning coal was suggested by Susan Leland.

This material is based on research supported in part by the U.S. Department of Energy under contract No. EX-76-A-01-2295 with MIT. Any opinions, findings, conclusions or recommendations contained herein are those of the authors and do not necessarily reflect the views of the U.S. Department of Energy.

PRINTED AT THE NIMROD PRESS, BOSTON

Library of Congress Cataloging in Publication Data

World Coal Study.

    Coal—bridge to the future.

    Bibliography: p.

    1. Coal.   I. Wilson, Carroll L.   II. Title.

TP325.W82   1980   333.8'22   80-13072

ISBN 0-88410-099-5

TABLE OF CONTENTS

FOREWORD

The World Coal Study (WOCOL) has been an international project involving over 80 people from 16 major coal-using and coal-producing countries. This is our Final Report: the product of 18 months of intensive work.

WOCOL came into being as a result of my belief that there was an urgent need to examine the role that coal might play in meeting world energy needs during the next 20 years. Such a choice is being forced upon us. Oil from OPEC countries can no longer be relied upon to provide expanding supplies of energy, even with rapidly rising prices. Neither can nuclear energy be planned on for rapid expansion worldwide until present uncertainties about it are resolved. Yet the world's energy needs will continue to grow, even with vigorous energy conservation programs and with optimistic rates of expansion in the use of solar energy.

Coal already supplies 25 percent of the world's energy, its reserves are vast, and it is relatively inexpensive. But no major attempt has previously been made to examine the needs for coal on a global scale, to match requirements of users with potential capacities of producers, to look at the markets coal might fill, or to examine the obstacles to a rapid expansion in coal use and how they might be overcome.

This experiment in international collaboration in WOCOL follows directly from the Workshop on Alternative Energy Strategies (WAES), which I directed. In WAES we concluded, in our report published in early 1977,[1] that the world might find its oil supplies being constrained by limitations on exports from OPEC countries at some time during the 1980s. This report was generally regarded at the time as being too pessimistic. In fact, it proved to be optimistic—it is now clear that we are already in that position in 1980.

WAES assembled a group of people who could engage in free-ranging discussion and who worked hard together and reached agreement on a plausible range of global energy futures. These people came from key positions in governments and private and public orga-

---

1. *ENERGY: Global Prospects 1985-2000* (New York: McGraw-Hill, 1977).

nizations in the world's major energy-using countries, as well as from some of the largest OPEC countries. They came however as individuals, each free to represent his own views rather than those of the organization with which he was affiliated.

We learned a great deal in WAES, not just about energy, but about how to work together as an international group. These lessons have been applied in WOCOL. I was responsible for the selection of the individuals taking part. I sought a mix of nationalities, a balance of public and private sector viewpoints, a common concern with energy policy, a knowledge of the coal or energy industry, and the ability to work effectively in a group effort. The total was no greater than the number of people who could, in dialogue around a table, function as a working group. The language of the Study was English.

Members of WOCOL came from 16 countries that use 75 percent of the world's energy. They produce and use about 60 percent of the world's coal. Although not able to take a full part in the Study, observers from the People's Republic of China attended WOCOL meetings and furnished a report for their country.

The senior members of WOCOL, called Participants, gave policy direction to the Study and reviewed its work as it proceeded. It is their consensus on the Study conclusions that represents the principal summary of WOCOL. Moreover, Participants chose Associates from within their organization who devoted much of their time to WOCOL under the direction of the Participant and collaborated with other Associates both in their own country and elsewhere. Each Participant arranged for the financing of all work done at a national level and for the time, travel, and other expenses of himself and his Associate.

Formal liaison was established with the U.S. Department of Energy and with the Electric Power Research Institute (EPRI), which both provided Study members who served in the role of Associates without Participants. Costs of the headquarters, located at the Massachusetts Institute of Technology, Cambridge, Massachusetts, were sponsored by foundations, companies, and the U.S. Department of Energy. A small, capable, and extremely dedicated staff at M.I.T. coordinated this global project within some very tight deadlines.

WOCOL Participants developed together an agreed purpose, plan, and schedule for the Study. Cases were developed to project a range of future coal production, use, and trade. Like WAES these projections build on detailed country studies conducted by WOCOL teams as well as on special studies for regions not represented in the Study. Unlike WAES the projections were based on existing national energy studies rather than on a specific global macroeconomic framework developed for the Study.

Our analysis goes to the year 2000. Over the next 5 years the use of different fuels is heavily constrained by decisions made in the past and by the fact that new energy facilities being planned now cannot be brought to completion in less than 5 to 10 years. The substantial shift to coal will therefore begin in the mid to late 1980s based on decisions being made in the early 1980s. Major effects of this expansion will be felt in the 1985–2000 period. The twenty years from now to the turn of the century are thus critical. The full effects of the switch to coal will, however, not be seen until the early decades of the next century. At that time the need for coal will still be great although renewable energy systems should be coming into use on a significant scale.

In the immediately following pages are the names and affiliations of those who took part in WOCOL; our conclusions; a description of how the Study actually proceeded; and acknowledgments of the support that made it possible. Part 1 of this volume, "The Need for Coal," is a summary of our report. It contains the essential message of the Study. Part 2, "Building the Bridge," is a more detailed presentation of the technical, economic, and logistical issues that have been considered in the Study and that support the findings of Part 1.

Volume 2, *Future Coal Prospects: Country and Regional Assessments,* which is separately published, contains the full texts of the reports by the teams from each country in WOCOL as well as the specially commissioned studies of other regions. These reports describe the levels of coal expansion that each country now projects may be required and the means to achieve them.

The WAES study left the world with a well-defined world energy problem. WOCOL presents a solution to that problem—*COAL —Bridge to the Future.* This is an optimistic message about energy— a rarity these days. But this optimism is justified only if many decisions are taken very soon because of the long lead times required. Our most precious resource is time which must be used as wisely as energy.

CARROLL WILSON

STUDY PARTICIPANTS, ASSOCIATES, AND STAFF

PROJECT DIRECTOR
Carroll L. Wilson

PARTICIPANTS

Australia
Dr. Donald D. Brown
Deputy General Manager and
    Director
CSR Limited

Dr. Robert A. Durie
Chief Scientist
R.W. Miller Co. Pty. Ltd.

Mr. Kenneth G. Wybrow
Member, Joint Coal Board

Canada
The Hon. Jack Austin Q.C.
The Senate

Mr. Garnet T. Page
President
The Coal Association of Canada

Dr. Charles H. Smith
Senior Assistant Deputy Minister
Department of Energy, Mines &
    Resources

China, Peoples Republic of
Mr. Liu Huanmin[1]
Director, China Coal Society
Advisor, China Coal Industry
    Technical Installation Corporation

Denmark
Mr. Poul Sachmann
Executive Vice President
ELSAM

Finland
Mr. Kaarlo K. Kirvelä
Vice President and Director
    of the Energy Sector
Ekono Oy

---

1 Observer

France
Dr. Thierry de Montbrial
Director, Institut Français
    des Relations Internationales
Professor, L'Ecole Polytechnique

M. Jean-Claude Sore
Deputy Managing Director
Charbonnages de France

M. Albert Viala
Vice President
A.T.I.C.

Germany, Federal Republic of
Dr. rer. oec. Harald B. Giesel
Member of the Board
Gesamtverband des deutschen
Steinkohlenbergbaus

Dr. H-D. Schilling
Head, Department of Coal Utilization
Bergbau-Forschung GmbH

Professor Hans K. Schneider
Director
Institute of Energy Economics
University of Koeln

India
Shri S.K. Bose
Joint Secretary
Ministry of Energy

Indonesia
Mr. S. Sigit
Secretary General
Department of Mines & Energy

Italy
Dr. Marcello Colitti
General Manager
ENI

Professor Umberto Colombo
Chairman, Comitato Nazionale per
    l'Energia Nucleare (CNEN)

Japan
Mr. Toyoaki Ikuta
President
Institute of Energy Economics

Mr. Hidezo Inaba
Vice Chairman
Committee for Energy Policy
  Promotion

Dr. Saburo Okita[2]
Chairman
Japan Economic Research Center

Mr. Setsuo Takagaki
Director of Research Affairs
Institute of Energy Economics

Netherlands

Dr. A.A.T. van Rhijn
Deputy Director-General for Energy
Ministry of Economic Affairs

Netherlands/UK

Mr. Frank Pecchioli
Managing Director
Shell Coal International Ltd.

Poland

Mr. Eugeniusz Ciszak
Deputy Managing Director,
Director Tech.
Chief Mining Studies & Design Office
(Ministry of Mining)

Sweden

Mr. Arne S. Lundberg
The Royal Swedish Academy of
  Engineering Sciences

United Kingdom

Mr. Robert Belgrave
Director, BP Trading Ltd.
Policy Adviser to Board of Directors,
  British Petroleum Co.

Sir Derek Ezra
Chairman
National Coal Board

Professor Sir William Hawthorne
Chairman, Advisory Council on
  Energy Conservation (1974-1979)
Master, Churchill College

Sir Ronald McIntosh
Executive Director
S.G. Warburg & Co. Ltd.
Former Director General of National
  Economic Development Office

United States

Mr. Thornton F. Bradshaw
President
Atlantic Richfield Co.

Mr. Gordon R. Corey
Vice Chairman
Commonwealth Edison Co.

Mr. W. Kenneth Davis
Vice President
Bechtel Power Corp.

Mr. Pierre Gousseland
Chairman and Chief Executive Officer
AMAX Inc.

Prof. Robert C. Seamans, Jr.
Dean, School of Engineering
Massachusetts Institute of Technology
Former Administrator, Energy
  Research & Development Admin.
  (ERDA)

Mr. Russell E. Train
President, World Wildlife Fund–U.S.
Former Administrator, Environmental
  Protection Agency (EPA)

**International Institute of
Applied Systems Analysis**

Prof. Wolf Häfele
Deputy Director
Program Leader, Energy Systems

ASSOCIATES

Australia

Mr. Keith Laverick[3]
Market Development Manager–Coal
The Broken Hill Proprietary
  Company Limited

Mr. David B. Tolmie
Principal Project Officer
New Business & Development Group
Minerals & Chemicals Division
CSR Limited

Canada

Ms. Barbara Ettles[4]
Manager of Information Services
The Coal Association of Canada

---

[2] Until appointed Foreign Minister November 1979

[3] Associate without Participant

[4] Until September 1979

Dr. John H. Walsh
Senior Advisor, Energy Technology
Dept. of Energy, Mines & Resources

China, Peoples Republic of[5]
Mr. Wang Quingyi
Engineer
China Coal Society

Mr. Li Weitung
Professor, Deputy Dean of the
    China Mining Institute

Ms. Sung Yafan
Engineer
China Coal Society
China Coal Industry Technical
    Installation Corporation

Mr. Li Zhongmin
Engineer
China Coal Society

Denmark
Mr. Hartmut Schulz
Manager Coal Purchase
ELSAM–Fuels Division

Finland
Dr. Seppo Hannus
Senior Science Officer
Ministry of Trade and Industry
Energy Department

Mr. Juha Kekkonen
Senior Advisor
Ministry of Trade and Industry
Energy Department

France
M. Daniel Cretin
A.T.I.C.

M. Marc Ippolito
Head of EEC and International
    Relations Department
Charbonnages de France

Germany, Federal Republic of
Dr. Dieter Schmitt
Head, Institute of Energy Economics
University of Koeln

Mr. Detlef Wiegand
Chief Economist
Dept. of Coal Utilization
Bergbau-Forschung GmbH

5 Observers

Italy
Dr. Oliviero Bernardini
Planning and Development/
    Technology Assessment
Montedison

Dr. Eugenio Nardelli
Division for Planning & Development
ENI

Japan
Mr. Seiko Ichikawa
General Manager of First Overseas
    Coal Dept.
Mitsui and Co. Ltd.

Mr. Tohru Kimura
Chief Economist
Institute of Energy Economics

Mr. Akira Kinoshita
Senior Economist
Dept. of General Planning
Electric Power Development
    Company

Mr. Seiji Kobayashi
Manager, Thermal Coal Section
C. Itoh & Co. Ltd.

Mr. Takao Sato
Manager, Research Division
Committee for Energy Policy
    Promotion

Netherlands
Mr. Rik Bosma
Manager of Coal Studies
Energy Study Center
Netherlands Energy Research
    Fnd/ECN

Mr. Robert van der Wart
Program Coordinator, Non-Nuclear
    Energy Systems
Netherlands Energy Research
    Fnd/ECN

Netherlands/UK
Mr. Brian K. Elms
Planning Manager
Shell Coal International Ltd.

Sweden
Mr. Kurt Lekås
Project Manager, Coal 90
Luossavaara-Kiirunavaara AB

Mr. Håkan Neuman
Economist, Corporation Development
Luossavaara-Kiirunavaara AB

United Kingdom

Mr. Peter R. Horrobin
Executive Director
S.G. Warburg Co. Ltd.

Dr. William Q. Limond
Policy Analyst, Policy Review Unit
British Petroleum Co. Ltd.

Mr. Kenneth N. McKinlay
Manager, Planning & Economics
BP Coal Department

Mr. Richard Ormerod
Department of Central Planning
National Coal Board

Mr. Michael Parker
Director of Central Planning
National Coal Board

United States

Dr. Arnold B. Baker
Senior Consultant
Policy Analysis & Forecasting
Atlantic Richfield Co.

Mr. Michael Gaffen[3]
Director, International Coal Analysis
U.S. Department of Energy

Dr. Irving Leibson
Vice President
Bechtel Inc.

Mr. Robert L. Major
Manager, Business Research
AMAX Coal Company

Mr. René H. Malès[3]
Director, Energy Analysis &
    Environment Division
Electric Power Research Institute

Mr. Joseph P. McCluskey
Director, Environmental Affairs
Commonwealth Edison Co.

Mr. F. Taylor Ostrander
Assistant to the Chairman
AMAX Inc.

Dr. John Stanley-Miller[3]
Director, Energy Modeling &
    Analysis Division
U.S. Department of Energy

Dr. David Sternlight
Chief Economist
Atlantic Richfield Co.

STAFF

MIT Program Staff

Mr. Robert P. Greene
Deputy Director

Dr. J. Michael Gallagher
Technical Director

Mr. Gerald Foley
Consultant

Mr. Ralph Chang
Research Assistant

MIT Support Staff

Ms. Roberta Ferland
Secretary

Ms. Susan Leland
Administrative Assistant

Ms. Lynette McLaughlin
Office Assistant

Ms. Kathleen Romano
Financial Manager

Ms. Susan Williamson
Administrative Secretary

European Support Staff

Ms. Karin Berntsen, Luxembourg

Ms. Elaine Goldberg, Spain

Ms. Dalia Jackbo, Norway

---

3 Associate without Participant

CONCLUSIONS

CONCLUSIONS

It is now widely agreed that the availability of oil in international trade is likely to diminish over the next two decades. Vigorous conservation, the development and rapid implementation of programs for nuclear power, natural gas, unconventional sources of oil and gas, solar energy, other renewable sources, and new technologies will not be sufficient to meet the growing energy needs of the world. A massive effort to expand facilities for the production, transport, and use of coal is urgently required to provide for even moderate economic growth in the world between now and the year 2000. Without such increases in coal the outlook is bleak.

Our major conclusions after eighteen months of study are as follows.

1. Coal is capable of supplying a high proportion of future energy needs. It now supplies more than 25% of the world's energy. Economically recoverable reserves are very large—many times those of oil and gas—and capable of meeting increasing demands well into the future.

2. Coal will have to supply between one-half and two-thirds of the additional energy needed by the world during the next 20 years, even under the moderate energy growth assumptions of this Study. To achieve this goal, world coal production will have to increase 2.5 to 3 times, and the world trade in steam coal will have to grow 10 to 15 times above 1979 levels.

3. Many individual decisions must be made along the chain from coal producer to consumer to ensure that the required amounts are available when needed. Delays at any point affect the entire chain. This emphasizes the need for prompt and related actions by consumers, producers, governments, and other public authorities.

4. Coal can be mined, moved, and used in most areas in ways that conform to high standards of health, safety, and environmental protection by the application of available technology and without unacceptable increases in cost. The present knowledge of possible carbon dioxide effects on climate does not justify delaying the expansion of coal use.

5. Coal is already competitive in many locations for the generation of electricity and in many industrial and other uses. It will extend further into these and other markets as oil prices rise.

6. The technology for mining, moving, and using coal is well established and steadily improving. Technological advances in combustion, gasification, and liquefaction will greatly widen the scope for the environmentally acceptable use of coal in the 1990s and beyond.

7. The amount of capital required to expand the production, transport, and user facilities to triple the use of coal is within the capacity of domestic and international capital markets, though difficulties in financing large coal projects in some developing countries may require special solutions.

The final conclusions of this Study are cautiously optimistic. Coal can provide the principal part of the additional energy needs of the next two decades. In filling this role it will act both as a bridge to the energy systems of the future and as a foundation for the continued part that coal will play in the next century. But the public and private enterprises concerned must act cooperatively and promptly, if this is to be achieved. Governments can help in particular, by providing the confidence and stability required for investment decisions, by eliminating delays in licensing and planning permissions, by establishing clear and stable environmental standards, and by facilitating the growth of free and competitive international trade. A recognition of the urgent need for coal and determined actions to make it available in time will ensure that the world will continue to obtain the energy it requires for its economic growth and development.

ACKNOWLEDGMENTS

Sponsors and Institutions

Many organizations and institutions have provided direct financial support and other services for WOCOL. The Study gratefully acknowledges the following sponsors for their generous support of the secretariat and related activities:

AMAX
Atlantic Richfield Company
Bechtel National Inc.
Cummins Engine Foundation
Alfried Krupp von Bohlen und
  Halbach Foundation

Massachusetts Institute of Technology
The Andrew W. Mellon Foundation
The Rockefeller Foundation
U.S. Department of Energy

The Study also gratefully acknowledges the many institutions in each country that have contributed financial or professional support for the Study or for the WOCOL country reports including:

Australia
The Broken Hill Propriety Company
  Limited
CSR Limited
Joint Coal Board
R.W. Miller Co. Pty. Ltd.

Commonwealth Government
  Department of National Develop-
    ment and Energy
  Department of Trade and Resources
  National Energy Advisory
    Committee

New South Wales Government
  Coal Export Strategy Study Task
    Force
  Department of Mineral Resources
    and Development
  Electricity Commission of N.S.W.
  Energy Authority of N.S.W.
  Hunter Valley Research Foundation
  Maritime Services Board
  Planning and Environment
    Commission
  Public Transport Commission of
    N.S.W.
  State Pollution Control Commission
New South Wales Combined Colliery
  Proprietors' Association

Queensland Government:
  Co-ordinator General's Department

State Electricity Commission of
  Queensland
Department of Mines
Queensland Colliery Proprietor's
  Association

Victorian Government:
  State Electricity Commission of
    Victoria
  Victorian Brown Coal Council

Canada
Department of Energy, Mines &
  Resources
The Coal Association of Canada

China, People's Republic of
China Coal Society

Denmark
ELSAM
Odense Steel Shipyard Ltd.

Finland
Ekono Oy
Ministry of Trade and Industry

France
Association Technique d'Importation
  Charbonnière (ATIC)
Charbonnages de France

ACKNOWLEDGMENTS

Institut Français des Relations
  Internationales (IFRI)

Germany, Federal Republic of

Bergbau-Forschung GmbH
Gesamtverband des deutschen
  Steinkohlenbergbaus
Institute of Energy Economics/
  University of Koeln

India

Central Mine Planning Design
  Institute/Ranchi Planning
  Commission

Indonesia

Department of Mines & Energy

Italy

Comitato Nazionale per l'Energia
  Nucleare (CNEN)
Ente Nazionale Idrocarburi (ENI)
Montedison

Japan

Committee for Energy Policy
  Promotion
C. Itoh and Co. LTD
Electric Power Development Company
Federation of Electric Power
  Companies
Institute of Energy Economics
Japan Economic Research Center
Mitsui and Co. LTD

Netherlands

Energie Studie Centrum ECN
Havenbedrijf der Gemeente Rotterdam
Havenbedrijven van Amsterdam,
  Delfzijl, Dordrecht en Terneuzen
Hoogovens IJmuiden B.V.

N.V. tot Keuring van
  Elektrotechnische Materialen
Ministerie van Economische Zaken
Nederlandse Energie Ontwikkelings
  Maatschappij B.V.
N.V. Nederlandse Gasunie
Shell Coal International Limited
Shell Nederland Verkoopmaatschappij
  B.V.
SHV Nederland N.V.
Stichting Energieonderzoek Centrum
  Nederland
Frans Swarttouw B.V.

Netherlands/U.K.

Shell Coal International Ltd.

Poland

Ministry of Mining/Warsaw

Sweden

Luossavaara-Kiirunavaara AB
The Swedish State Power Board
Sydkraft AB

United Kingdom

British Petroleum Company
National Coal Board
University of Cambridge:
  Engineering Department
S. G. Warburg & Co. Ltd.

United States

AMAX Inc.
Atlantic Richfield Co.
Bechtel Incorporated, and
  Bechtel Power Corporation
Commonwealth Edison Co.
Electric Power Research Institute
Massachusetts Institute of Technology
U.S. Department of Energy
World Wildlife Fund–U.S.

THE WORLD COAL STUDY

The WOCOL Approach

The Participants adopted a purpose, plan and schedule for the World Coal Study at their first meeting held at Aspen, Colorado, in October 1978. It was agreed that:

> "The World Coal Study is designed to be an action-oriented assessment of future prospects for coal. . . . Its objective is to examine the future needs for coal in the total energy system and to assess the prospects for expanding world coal production, utilization, and trade to meet these needs. . . . It will rely as much as possible on available and appropriate analysis performed by others. It will apply its own resources in areas where other satisfactory work is not available and it will undertake its own evaluation of possible coal development strategies. Environmental issues will be given special attention because of their importance in the expansion of the production and use of coal."

The Study has proceeded in accordance with these guidelines.

WOCOL Meetings

There were five week-long meetings of Study members from October 1978 through January 1980—three in Europe and two in the United States. Participants attended these meetings for the first two and one half days to review the progress of the work and to provide guidelines for the next phase of the Study. Associates met for the remainder of the week to formulate a work program which was then carried out by country teams, the coordinating secretariat at M.I.T., and groups of Associates with particular interest in and knowledge of a subject.

Each country team also met several times during the course of the Study to develop and review the country reports prepared for WOCOL.

Units

Throughout this report we use *mtce* or *million metric tons of coal equivalent* as our standard measure of energy. On an energy basis, 76 mtce/year is equivalent to 1 million barrels per day (mbd) of oil.

A *ton of coal equivalent* (tce) as used in this report is a metric ton (2,205 pounds) of coal with a specific heating value (7,000 kcal/kg or 12,600 Btu/lb). Because coals vary significantly in heat content, more than 1 metric ton of coal is often required to produce the energy content of 1 tce. For example 1 tce is equivalent to 1.4 metric tons of subbituminous western United States coal (assuming 9,000 Btu/lb).

Uncertainties

The fragility and uncertainty of the world's energy system has become increasingly clear during the 18 months of this Study. The world price of oil increased from $13/barrel to $30/barrel during this time. The developments in Iran and Afghanistan and the Three Mile Island nuclear accident have all had their impacts, the full effects of which remain to be seen. They provide examples of what may lie ahead.

There could also be pleasant surprises. Technological break-throughs in a variety of areas, such as more economical or more efficient processes for obtaining oil from shale, tar sands, or heavy oils; or an increase in the average recovery rate of conventional oil; or in-situ coal gasification; or the extraction of gas from tight forma-tions; or the rapid development and deployment of low-cost solar energy technologies could all make more energy available than a prudent estimate of energy prospects would now consider likely.

The WOCOL conclusions are based on the facts available. The members of the Study have built into projections for their coun-tries expected changes in government policy, particularly those relat-ing to a more rapid increase in energy efficiency (conservation). But the analysis cannot include the effects of surprises. Surprises are just that—unpredictable events.

Study Reports

This Report, *COAL—Bridge to the Future,* is the Final Re-port of the World Coal Study. It represents the collective effort of the Study members. It is the result of our individual work in each country and our deliberations together, which included 5 week-long meetings during the 18 months of the project. It is accompanied by a second volume—*Future Coal Prospects: Country and Regional Assessments* —containing the full text of the comprehensive country studies by WOCOL teams in the 16 countries participating in the Study as well as assessments for other regions of the world.

This Report

Most of the substance of this Report was completed on January 18, 1980. Data collection by WOCOL members began in October 1978 and was largely complete by October 1979.

Many authors contributed to the Report, and as a result there is some variety in the prose. Editing has not aimed at eliminating this completely, but has concentrated on consistency in terminology, units, and the like.

Participants and Associates have taken part in WOCOL as individuals and this report therefore does not necessarily represent the views of the public or private organizations with which they are associated. No single member had the time or the expertise to judge every topic covered; moreover, on some topics individuals held different views. The Report in such cases has sought the consensus of the majority and it is accepted that each member does not necessarily subscribe to all the statements of the Report. Nevertheless, all WOCOL members agree on the general analysis and the main findings of the Report; and all believe it will contribute to a better understanding of the role coal must play in building the energy bridge to the future.

The Conclusions were discussed in detail by Participants at the final WOCOL meeting and the text included here is as agreed at that meeting.

Part I of this Report, "The Need for Coal," is the principal summary report of the Study Participants. Gerald Foley prepared the first two drafts of this summary and helped to structure the entire Report. The last two drafts were extensively reviewed and revised by all Study members. I am responsible for the final editing of Part I based on guidance given by Participants at our final meeting.

Part II of this Report, "Building the Bridge," is also a product of the entire Study. It differs from Part I because one member of the Study took principal responsibility for preparing the individual chapters with technical and editorial assistance from other Study members.

Chapter 1, World Energy Prospects, was written by J. Michael Gallagher. Kenneth McKinlay and Robert van der Wart assisted with final drafting.

Chapter 2, Analysis of World Coal Prospects, was written by J. Michael Gallagher. Oliviero Bernardini assisted with final drafting.

Chapter 3, Coal Markets and Prices, was written by Arnold Baker. Dieter Schmitt assisted with final drafting.

Chapter 4, Environment, Health and Safety, was written by René Malès with assistance from Robert Greene. Akira Kinoshita assisted with final drafting.

Chapter 5, Coal Resources, Reserves and Production, was written by Robert Major with assistance from John Walsh and Detlef Wiegand.

Chapter 6, Maritime Transportation and Ports, is based on technical studies prepared by Per Bech of Odense Steel Shipyard Ltd. and by Shell Coal International Limited and was jointly drafted by Brian Elms, Eugenio Nardelli, and David Tolmie.

Chapter 7, Coal-Using Technologies, is based on several drafts prepared by Detlef Wiegand and was jointly drafted by Robert Durie and Irving Leibson with assistance from H.D. Schilling and Detlef Wiegand.

Chapter 8, Capital Investment in Coal, was written by Carroll Wilson with assistance from Peter Horrobin, Richard Ormerod, and F. Taylor Ostrander.

Each of the chapters was reviewed by all Associates and some Participants at various stages of drafting. Gerald Foley and Katharine Parker provided editorial and copy editing assistance. Diane Leonard-Senge and Norma Wilton prepared the visuals. Walter Tower and Norma Wilton designed the book and cover.

Robert P. Greene and J. Michael Gallagher managed this Report and the project secretariat assisted by Susan Leland, Susan Williamson, and Kathleen Romano. Coordinating such a project of 80 people from 16 countries, bringing it to an agreed conclusion in 16 months, and producing a 2-volume report against very tight deadlines has taken exceptional talents, energy, and teamwork. At M.I.T. they have been aided by Ralph Chang, Roberta Ferland and Lynette

McLaughlin; at our meetings in Europe, by Karin Berntsen, Elaine Goldberg, and Dalia Jackbo.

Other Acknowledgments

WOCOL gratefully acknowledges the assistance provided by the International Energy Agency (IEA) primarily through its report *Steam Coal Prospects to 2000,* which was published in December 1978. Mr. John R. Brodman of IEA's Energy Economics Division was particularly helpful in a liaison role with WOCOL at the beginning of this Study.

Mr. H. T. Burger of the Department of Environmental Planning and Energy of the Republic of South Africa provided projections for that country.

Mr. Per Bech, Assistant Director, Odense Steel Shipyard Ltd., of Denmark provided assistance with the Maritime Transport study.

Other contributors were:

Mr. Lisle Baker
Professor of Law/Suffolk University
Boston, Massachusetts

Mr. George Curtin
Production, Nimrod Press
Boston, Massachusetts

Mr. Gerald Foley
Editor/Writer
IIED/London

Ms. Diane Leonard-Senge
Graphic Artist
Cambridge, Massachusetts

Mr. Osgood Nichols
Osgood Nichols Associates, Inc.
Wilton, Connecticut

Ms. Katharine Parker
Copyeditor
Cambridge, Massachusetts

Mr. Walter Tower
President, Nimrod Press
Boston, Massachusetts

Mrs. Norma Wilton
Art Director, Nimrod Press
Boston, Massachusetts

On behalf of the members of the World Coal Study I wish to express our great appreciation to all who have helped to make this report possible.

C. L. W.

THE NEED FOR COAL
(Summary)

THE NEED FOR COAL

Introduction — World Energy Prospects — Prospects for Coal — Coal Markets and Prices — Environment, Health, and Safety — Coal Resources, Reserves, and Production — Ports and Maritime Transport — Technologies for Using Coal — Capital Investment in Coal — Necessary Actions

Introduction

The world has reached the end of an era in its energy history. Increased supplies of oil, which have been the basis for economic growth in the past few decades, are not expected to be available in the future. The development of a new energy basis for continued economic growth has therefore become an urgent necessity. Building a bridge to the energy sources and supply systems of the next century —whatever they turn out to be—is of crucial importance. We believe that coal can be such a bridge and that it will also continue to serve a vital role into the longer-term future.

A tripling of coal use and a 10–15 fold increase in world steam coal trade would allow the energy problems of the next two decades to be faced with confidence. With such increases in coal use, coupled with a mobilization of other energy sources and vigorously promoted conservation, it becomes possible to see how to meet the energy requirements of moderate economic growth throughout the world. But without such a coal expansion the outlook is bleak. This is the central message of our report.

In the industrialized countries coal can become the principal fuel for economic growth and the major replacement for oil in many uses. Coal may also provide the only way for many of the less developed countries to obtain the fuel needed for electric power and industrial development, and to reduce their dependence upon oil imports.

Increased use of coal will require large investments, but no greater than those required for other fuels such as oil, gas, and nuclear energy. Countries now heavily dependent upon oil must build coal-using facilities on a large scale before they can use the coal they will need. Power stations, coal ports, railways, and handling facilities take a long time to plan and build. So do mines and export terminals. Unless decisions to build them are made soon, these facilities will not be ready by the time when this study indicates that they will be acutely needed. It is necessary for governments and industry to act cooperatively so that the required investment decisions are made promptly.

Unlike oil, the reserve base for coal is sufficiently great to support large increases in production for a long time into the future. Moreover, the technology for its safe and environmentally acceptable production, transport, and use is proved and already widely applied in most areas.

We have examined environmental questions with great care and considered the measures that would have to be taken within each WOCOL country to comply with present and anticipated environmental regulations. The technology exists by which exacting standards of environmental protection can be met, and much work is being done to improve it and lower its costs. We are convinced that coal can be mined, moved, and used at most locations in ways that meet high standards of environmental protection at costs that still leave coal competitive with oil at mid-1979 prices.

The present knowledge of the effects on climate of carbon dioxide (CO_2) from fossil fuel combustion does not warrant a global reduction in fossil fuel use or a delay in the expansion of coal. However, support for research on the possible effects of increased atmospheric CO_2, including that of global warming, is essential so that future policies may reflect an improved understanding of such matters.

Coal is not in competition with conservation, nuclear or solar energy, or other sources as the sole solution to the world's energy problems. All these will be required if the energy needed is to be supplied. The world, however, needs an incremental energy source as nearly like oil as possible, but with the vital difference that it will be obtainable in increasing amounts until well into the next century. Ideally, and if it is to fill the role played by oil over the past decades, it should be versatile in application, easily transported and stored, and reasonably priced. The technology for using it should be mature and generally available so that it can be brought into use rapidly, widely, and safely. It should be capable of satisfying strict environ-

4

mental standards with presently available technology and at a cost competitive with other fuels. It should be obtainable in large quantities and for long enough to justify the investments required to bring it into widespread use. Only coal comes close to meeting these specifications.

This need for coal has been recognized many times, perhaps most clearly and conclusively in the resolution adopted at the Seven-Nation Economic Summit at Tokyo in June 1979:

> We pledge our countries to increase as far as possible, coal use, production, and trade, without damage to the environment. We will endeavor to substitute coal for oil in the industrial and electrical sectors, encourage the improvement of coal transport, maintain positive attitudes toward investment for coal projects, pledge not to interrupt coal trade under long-term contracts unless required to do so by a national emergency, and maintain, by measures which do not obstruct coal imports, those levels of domestic coal production which are desirable for reasons of energy, regional and social policy.[1]

The International Energy Agency has also been strongly supportive of greater coal use and has adopted a statement of "Principles for IEA Action on Coal," agreed upon steps designed to increase the use of coal, and has created a high-level Coal Industry Advisory Board (CIAB) to assist in the practical implementation of these principles.

Declarations of intent are, however, not self-executing. They require action. We believe that a deeper public understanding of the importance of coal to the world's future is necessary if the numerous steps by local, regional, and national governments, by international agencies, and by public and private investors are to be taken in time to allow expanded use of coal to provide a major part of the energy required for the world's continued economic growth.

World Energy Prospects

Worsening Outlook for Oil

During the past two decades world oil consumption has grown twice as fast as that of all other sources combined. Two-thirds of

---

1. Abstract from Tokyo Communiqué of Seven-Nation Economic Summit, June 1979. Countries included Canada, France, the Federal Republic of Germany, Italy, Japan, the United Kingdom, and the United States.

the growth in energy consumption in the member countries of the Organization for Economic Cooperation and Development (OECD) came from oil during this time. In Western Europe and Japan, in fact, oil has provided over 80 percent of this increase in energy use. In all but a few of the developing countries oil has also supplied virtually all of the increase in energy consumption. It is a matter of considerable concern therefore that there now appears to be no realistic prospect of oil meeting any substantial part of future increases in the world's energy needs.

The present international oil trade amounts to about 35 million barrels a day (mbd), which is about 55 percent of the world's total oil consumption. Of that amount about 80 percent is exported by the OPEC countries, and more than 20 mbd flows through the Straits of Hormuz on the Persian Gulf. It is this concentrated and vulnerable flow of oil from the OPEC countries that has provided virtually all the flexibility within the world's energy system since 1960.

There are no secure grounds for assuming that OPEC production of oil will increase in the future, and there are good reasons to fear it may be less. As the Iranian revolution has shown, political change can be rapid and unexpected; and there are many other tensions, particularly in the Middle East, that are unlikely to be resolved soon.

Moreover, key producers, such as Saudi Arabia and Kuwait, have adopted policies that will conserve oil in the ground as a better security for their future than income they cannot spend or usefully invest. It is now the declared policy of OPEC member countries to limit oil production to amounts that will total not more than the present level of about 30 mbd. This policy is directed toward optimizing rather than maximizing revenues. The goal is to generate only as much revenue as their economies can handle without causing too rapid social change and high inflation. Increasing prices while holding production stable, or even reducing it, meets this financial objective and at the same time conserves oil reserves.

There is nothing unexpected or mysterious about this. The signs have been visible for some time. In 1977, the WAES report, *Energy: Global Prospects 1985–2000*, the direct predecessor of this Study, pointed to the problems likely to arise from a decision by OPEC members to restrict oil production in the early 1980s. It said that

6

Such an early limit could impose severe strains on all consuming countries. It is almost certain that the resulting shortfall in energy supply would lead to a significant price rise. This might trigger a world recession that would increase unemployment in many or all industrialized countries and severely damage the economies of the developing countries.

This is what is happening, but it is occurring a few years sooner and with a production ceiling lower than expected.

Prospects for increased oil supplies from non-OPEC areas are not sufficiently encouraging to change this picture. Production from fields in the North Sea and Alaska's Prudhoe Bay are projected to peak and level off in the next few years. It appears that Mexico and Norway will continue to limit exports to meet revenue needs. Increases in oil production in the People's Republic of China and the developing countries will be largely used to meet growing domestic demand. Furthermore some experts have suggested that production in the Soviet Union, which now exports 1 mbd to countries outside Eastern Europe, is leveling off, and that the Soviet Union/Eastern Europe area is therefore unlikely to remain self-sufficient in oil.

For all these reasons, we believe that oil will provide, at most, a small fraction of any future increase in world energy needs. Indeed, the availability of oil for import by the OECD nations is likely to be less in the year 2000 than today.

The specific amount of oil available to OECD member countries will depend on a number of factors. There are a range of projections, which we discuss in detail in Chapter 1, about production and domestic consumption of oil by OPEC member countries, oil exports from other countries, and oil import requirements of other regions including the developing countries and the centrally planned economy countries. These assumptions lead to a projected range of net oil imports available to the OECD which we have illustrated in Figure I-1. The most likely projection within that range is for a gradual decline in imports from 26 mbd in 1978 to about 22 mbd by the year 2000. The world must find ways other than oil to provide the additional energy required for its future economic growth.

Natural Gas and LNG

Natural gas provides a substitute for many uses of oil. World reserves are as great as those for oil, whereas use is at present much

7

Figure I-1   Range of Net Oil Imports Available to OECD Countries (1960–2000)

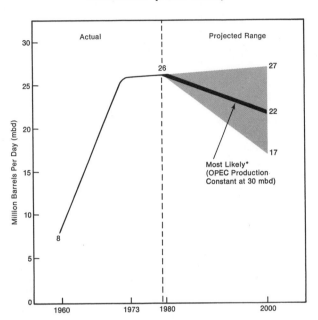

* Corresponds to assumptions of (1) constant OPEC oil production of 30 mbd, (2) an increase in OPEC internal oil consumption from 2.2 mbd in 1978 to 5 mbd in 2000, (3) a decrease in net CPE oil exports from 1.0 mbd to 0 mbd from 1978 to 2000, and (4) constant net oil imports of 3.2 mbd from 1978–2000 for other countries outside the OECD. Other combinations of assumptions could yield similar results on net oil imports available to OECD countries.

less. But in some areas where natural gas is already produced and used on a large scale, such as the United States and parts of Western Europe, expansion is already being curtailed as consumption exceeds new discoveries. Imports of gas by pipeline or as liquefied natural gas (LNG) are significant in some countries and may increase. But the rate of increase is likely to be slow because of the long lead times and the large capital investment required for undersea pipelines, liquefaction, transport, and regasification facilities.

Moreover, any substantial increase in the world use of natural gas would depend on large exports from countries in the Middle East and North Africa, which are members of OPEC. If, as now appears likely, these countries impose ceilings on their natural gas exports similar to those now being imposed on oil exports, major increases in world natural gas use would not occur, and natural gas prices could tend to closely reflect world oil prices.

8

Unconventional Oil and Gas Sources

In several countries, a vast resource base exists for unconventional oil in the form of oil shale, tar sands, and heavy oil. Canada for example already produces about 7 percent of its oil from Alberta tar sand deposits. Coal beds, shales, tight sands, and geopressurized formations may also contain huge amounts of unconventional gas.

A substantial expansion of unconventional oil or gas production is now being planned for some countries. However, all these unconventional sources are characterized by high capital costs and long lead times for development, and their largest contributions are therefore likely to occur after the year 2000.

Nuclear Uncertainties

Nuclear power already plays a substantial role in electricity generation in some countries. In some regions of the United States and Europe, for example, it supplies up to 30 percent of the total electricity. In several countries it is expected to provide a large share of future needs.

However, nuclear energy has suffered well-publicized setbacks and disappointments in recent years. For example, OECD forecasts of installed capacity in 1985 have been reduced by 60 percent in the past 10 years; projections to the year 2000 have been reduced even more. Although these nuclear forecasts have been revised steadily downward, many of the earlier forecasts were based on very optimistic projections and the more recent estimates are thought to be more realistic.

There are still, however, political uncertainties in a number of countries about whether or not to expand the use of nuclear power. Until these issues have been resolved and public confidence restored, it is difficult to make confident predictions about the future contribution that nuclear power will make. Because the choice for new power plants is now almost entirely between coal and nuclear energy, any delay in realizing the currently projected nuclear programs will increase the requirement for coal. This does not mean that coal and nuclear power are necessarily in competition. To the extent that there are clear advantages in establishing a diversified basis for the production of electric power, the relationship between coal and nuclear energy will tend to be complementary rather than competitive.

9

Conservation

Aggressive conservation programs now occupy a central place in the energy strategies of many nations. It has been widely suggested that by the year 2000 conservation could hold world energy consumption to a level 20–30 percent below what it would otherwise have reached. Even greater savings are suggested in some projections. Over the next 20 years, conservation may well become one of the world's largest energy "sources."

The first 10–15 percent of energy savings are obtainable by relatively simple short-term measures such as good housekeeping, better insulation, and improved heating controls. In some sectors and in some countries where energy has been cheap, even larger short-term gains are possible.

To go beyond these immediate measures, however, capital must be invested, and lead times become longer. For instance, replacement of a country's industrial equipment stock may take 20 to 30 years and the turnover of its automobile fleet 7 to 10 years. The fact that many energy conservation decisions must be spread over the populace at large is an additional delaying factor. The achievement of substantial savings therefore tends to be gradual, even with government intervention in the form of tax and financial incentives, as well as regulations governing energy performance standards of buildings and equipment.

The projections made in the WOCOL country studies nevertheless assume that high levels of energy conservation will, in fact, be achieved. The assumed savings lead to about a 25 percent reduction by the year 2000 in the amount of energy used in the OECD per unit of economic activity (GNP). We strongly support vigorous energy conservation efforts to ensure that the conservation savings achieved are as large as possible; however, we conclude that conservation savings are not of themselves capable of eliminating the world's need for additional energy supplies.

Renewable Energy Sources

Hydroelectricity is presently the only renewable energy source that makes a substantial contribution to global energy needs. In the future its greatest expansion is likely to be in the developing countries, some of which have large untapped hydro resources. It can be expected to continue to provide about 5 percent of the world's energy requirements.

Geothermal sources provide useful quantities of energy in some areas today and additional sources are being discovered and developed. The total contribution from geothermal sources will nevertheless inevitably remain geographically limited and relatively small in relation to world energy needs.

Conversion of biomass to fuel alcohol is also under way in some countries, and it is already making a useful contribution to transport fuel requirements in Brazil.

Solar energy has probably the greatest long-term potential of the renewable resources, but its rate of expansion will be limited by the time required for new technologies to become economic and then penetrate energy markets. Solar collectors for water heating, and in some cases for space heating, are already commercially available. It has been suggested that about 2 percent, or perhaps somewhat more, of the world's primary energy consumption could be supplied in this way by the year 2000. Solar generation of electricity, whether by thermal methods or by photovoltaics, is not yet economic and is therefore unlikely to have a widespread commercial impact during the next 20 years.

Considerable work is also being carried out on many other renewable energy sources including wind, wave, and tidal power; ocean thermal gradients; and various methods of conversion of organic materials. None of these, at present, seems capable of making more than a small contribution to total energy needs by the year 2000, though some may have substantial local impacts.

Prospects for Coal

The dwindling prospects for any substantial increase in the supply of oil at acceptable prices constitutes the main reason for the increased importance of coal. Even with the most optimistic forecasts for the expansion of nuclear power and the aggressive development of all other energy sources, as well as vigorous conservation, it is clear that coal has a vitally important part to play in the world's energy future. World coal production in 1977 was about 3400 million metric tons of raw coal. This was a contribution of about 2500 million metric tons of coal equivalent (mtce) or 33 mbdoe,[2] already greater than any energy source except oil.

Projecting Coal Requirements

Our analysis of future world coal requirements is built upon detailed country studies conducted by WOCOL teams as well as special studies for regions not directly represented in this Study. The teams developed two reference cases to project their expected range of future coal demand. Case A considers a moderate increase in coal demand to year 2000, whereas Case B assumes a high increase in coal demand, one which would increase world coal supply, trade, and use to what now appear to be close to the feasible upper limits.

Each country team formulated its case estimates of moderate and high increases in coal use within the general economic, technological and political environment expected within its country. No attempt was made to develop, for these cases, a specific concensus about the general global economic framework over the next 20 years. The WOCOL country studies were based instead on readily available and detailed national energy studies modified to reflect the specific WOCOL focus on projecting a range of future coal requirements. There is considerable richness and diversity in these WOCOL country studies,[3] which describe the coal expansion countries now expect may be required to meet their future energy needs.

Projections of total coal use were developed by market sector in each country. Teams also provided estimates of indigenous coal production and the coal imports which might be required. The feasibility of satisfying coal import requirements was assessed by matching them against the estimates of coal export potential made by the major coal-producing countries, most of whom are represented in WOCOL.

The average growth in total energy consumption was lower in both cases than that used in some other recent energy studies, for example that of the IEA Steam Coal Study (December, 1978). The average growth in OECD energy use in WOCOL Case A was 1.75

---

2. 1 mbdoe is equivalent to 76 mtce/yr.
3. Details of the methods and assumptions used, as well as the results of the projections made by the WOCOL country teams, are described fully in Chapter 2. Appendix 1 provides a summary of the detailed data supporting the analysis for each country. The full country reports are contained in Volume 2, *Future Coal Prospects: Country and Regional Assessments*.

percent per annum for 1977–2000; that of Case B was 2.5 percent.[4] These compare with 2.7 percent per annum for the same period in the IEA Study.

Coal Use by Market Sector—OECD

The estimates of total coal use in Cases A and B were developed through a detailed analysis by market sector (electricity, metallurgical, industry, residential/commercial, and synthetic fuel) within each country.

The major coal use in the year 2000, as today, is projected to be in electricity generation, which consumes more than 60 percent of the total coal. Estimates of coal requirements in the electric market are strongly influenced by assumptions about rates of growth of electricity demand and the expansion of nuclear power, as well as the replacement rate of existing oil-fired capacity. Our analysis indicates that, even under the moderate electricity growth assumptions of Case A (3 percent/yr) coal-fired electric capacity in the OECD will need to more than double from 350 GWe[5] in 1977 to 825 GWe by year 2000. A 1 percent per year higher electricity growth rate (Case B) leads to a further 500 GWe increase in total electric capacity requirements, about half of which is projected in the WOCOL team reports to be coal-fired. This yields a coal-fired capacity in the OECD of 1090 GWe in year 2000 for Case B. The corresponding projections for coal requirements for the OECD electric market are an increase from 600 mtce in 1977 to year 2000 estimates of 1325 mtce in Case A and 1850 mtce in Case B.

The estimates reflect an increase in the percentage of OECD coal-fired electric capacity, from 32 percent in 1977 to about 40 percent in 2000. Significant regional variations continue in year 2000 with coal penetration in the electricity sector being about 85 percent in Australia, 50 percent in North America, 35 percent in Europe, and 15 percent in Japan.

Coal for metallurgical purposes is the second largest market for coal today in the OECD, accounting for about 250 mtce/yr, or 25

---

4. OECD growth rates resulting from WOCOL analyses refer to overall averages for the OECD as derived from individual country projections in the WOCOL studies.
5. Each GWe of electric capacity requires 2 mtce/yr coal if the plant is operated at a 65% capacity factor.

percent of total coal use. Its growth since 1960, when the total was 190 mtce, has been slow. The WOCOL projections show a moderate increase to 330-375 mtce/yr by the end of the century. However the metallurgical coal share of the total coal market is projected to fall to about 15 percent of total coal use by year 2000.

The use of coal in industry has declined rapidly over the past two decades and now accounts for only about 90 mtce, or 9 percent of total OECD coal use. This trend is expected to be reversed. In our projections, industrial coal use in the OECD countries expands by two to four times by year 2000. The major expansion is expected to occur after 1985 when the market is projected to grow at 5–7 percent a year. Industries where coal use is expected to increase significantly include cement, chemicals (as fuel and feedstocks), petroleum refining, and paper.

A substantial new market for coal as feedstock for synthetic oil and gas plants could develop in the 1990s. This appears to be particularly likely in the United States. The total OECD market for the coal required for synthetic fuels by year 2000 is estimated to range from a low of 75 mtce in Case A to a high of 335 mtce in Case B. The higher amount of coal would supply 65 large synfuel plants each producing about 50,000 bdoe of synthetic oil or gas.

Coal has nearly disappeared as a fuel for homes and commercial facilities in the OECD countries, with the use in 1977 being less than 50 mtce. This compares with five times as much, or nearly 250 mtce, as recently as 1960. Although WOCOL country studies suggest that the total residential/commercial use of coal may remain insignificant during this century, there are in some countries indications of a revival of interest in coal for use in homes and office buildings. Projections for the United Kingdom, for example, indicate that such coal use could grow from 10 mtce in 1985 to 15–21 mtce by the year 2000. Substantial growth in the use of coal for fuel in district heating plants is also projected in some countries, for example in Denmark.

Total Coal Use—OECD

The projected total coal use in the OECD, by country, as summed from the WOCOL market sector projections, is given in Table I-1. The projected growth in OECD coal requirements from 1977 to 2000 is substantial, about 1000 mtce/yr, or a doubling, in Case A; and about 2000 mtce/yr, or a tripling, in Case B. In a number of countries, for example Australia, Canada, Japan, and the United States, the increase is particularly large. The OECD demand for coal is projected to accelerate sharply after 1985 in the Case B projections.

Table I-1 Total Coal Requirements in OECD Countries (mtce)

Country/Region	1977	TOTAL[a] COAL			
		1985		2000	
		Case A	Case B	Case A	Case B
Canada	25	44	41	82	121
United States	509	655	725	1,075	1,700
Denmark	4.6	10.7	11.1	9.4	20.9
Finland	4.3	4.4	4.4	9	13
France	45	35	59	48	125
Germany, Fed. Rep.	102	119	126	150	175
Italy	13.5	22.3	22.9	31.5	60.5
Netherlands	4.5	10.4	10.4	23	38
Sweden	2.1	5.1	5.4	17	26
United Kingdom	109	107	115	133	179
Other Western Europe	51	61	83	135	175
Japan	79	97	102	150	224
Australia	38	65	65	141	166
Total OECD[b]	990	1,235	1,370	2,000	3,025

[a] Total includes steam plus metallurgical coal.
[b] Totals are rounded.

The Effect on OECD Coal Needs of Limitations on Oil Imports and Nuclear Delays

Our review of the projections in Cases A and B for total OECD oil imports and nuclear power expansion, summed from the individual country estimates, led to the formulation and examination of two additional cases.

Projected oil imports in the OECD increased in both Case A and Case B by about 3 mbd by year 2000, which implied that oil available for OECD countries to import would need to increase from 26 mbd in 1977 to 29 mbd in 2000. This exceeded our most optimistic estimate of the amount of oil that would be available. Case A-1 was therefore developed to be compatible with the assumption that OPEC oil production does not rise above today's levels, and that about 22 mbd would be available for import to OECD countries in the year 2000 (see Figure I-1). This required a 20 percent reduction in the total

15

use of oil projected within the OECD countries in year 2000 in Case A.

A 7–10 fold increase in nuclear power capacity in the OECD was also estimated in the reference projections: from about 80 GWe in 1977 to 550 GWe in the year 2000 in Case A, and to 775 GWe in Case B. Case A-2 was established to investigate the effects of possible delays in the expansion of nuclear power. The figure of a 30 percent reduction in the total OECD nuclear capacity projected in year 2000 in Case A was selected merely to illustrate the magnitude of the effects of possible nuclear delays. The Case A-2 nuclear capacity projected in the year 2000 was thus about 400 GWe and compares with the 260 GWe now operating or under construction in the OECD countries.

The effects of possible oil limitations and nuclear delays are reported for Case A only since the coal projections in Case B were already considered near the upper limits of what is considered plausible. Figure I-2 shows the effects of the two variations of Case A on OECD coal consumption. Case A-1 superimposes on Case A an incremental OECD coal demand of 500 mtce in the year 2000, calcu-

Figure 1-2 The Effects of Oil Limitations and Nuclear Delays on OECD Coal Requirements (1960–2000)

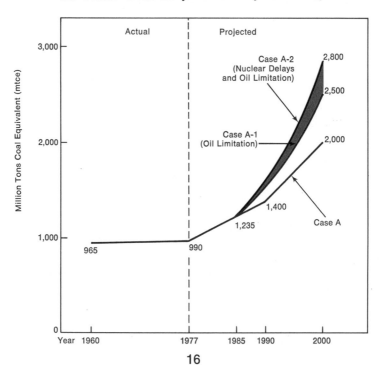

lated by the WOCOL teams as part of their responses to the oil limitation. In Case A-2 there is a further increase of 300 mtce/yr, calculated under the simplifying assumption that coal is substituted directly for the 30 percent nuclear reduction, which brings the total OECD coal use in Case A-2 to 2800 mtce in the year 2000.

Our assessment is that coal will be required to meet half or more of the energy increase in the OECD over the next two decades. Figure I-3 makes the point vividly. In the past two decades 67 percent of the total increase in OECD energy use was met by oil. Coal contributed practically nothing to the energy growth (Figure I-3a). The progressively increasing share of coal in meeting future energy growth in the Case A, A-1, and A-2 analyses is shown in the other diagrams with the maximum being 67 percent in Case A-2—exactly the same proportion of the increase taken up by oil in the past.

Figure I-3   Coal's Share in Meeting the Increase in OECD Energy Needs (1978–2000)

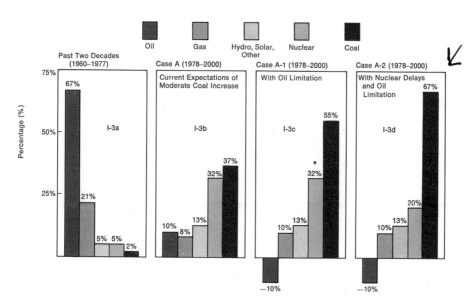

* In some countries the shortfall in oil imports in Case A-1 was assumed to be partly offset through increases in nuclear power supplies, but the effect on the OECD total was less than 1% as shown in Fig. I-3c.

Coal Use—Outside OECD

Countries outside the OECD used about 1500 mtce coal in 1977 or 50 percent more than OECD countries. The largest coal consumers were the Soviet Union (490 mtce), the People's Republic of China (368 mtce), Poland (127 mtce), India (72 mtce) and South Africa (61 mtce). While the available data does not allow us to make a precise analysis of future coal needs or to develop comprehensive Case A and Case B projections in all areas, it is apparent that these countries will be subject to many of the same pressures on their energy supplies as OECD countries.

The WOCOL analyses of developing country regions of East and Southeast Asia, Latin America, and Africa, together with the country team estimates for India and Indonesia, indicate a large increase in coal use in the developing countries, from 150 mtce in 1977 to a range of 600–900 mtce by the end of the century. India alone projects an increase from 72 mtce in 1977 to 280 mtce by the year 2000. Indonesia indicates a growth from less than 1 mtce in 1977 to about 20 mtce in the year 2000.

One of the principal coal uses will be for electricity generation in newly industrializing countries without large hydroelectric resources (e.g., South Korea, Taiwan, Philippines, and some countries of Latin America). Steam coal requirements in the developing countries of East and Southeast Asia are projected to be at least 70 mtce in year 2000, and could be as high as 190 mtce, compared to less than 20 mtce in 1977.

Coal use in the USSR and East European countries is projected to increase to about 1500–2000 mtce by the year 2000. The Poland study for WOCOL estimated that its domestic coal use will increase from 127 mtce in 1977 to 262 mtce in the year 2000. In the report from the People's Republic of China, coal use is projected to increase by three to four times, to well over 1000 mtce/yr.

Combining the WOCOL projections of OECD coal use with our more tentative projections for other regions indicates an increase in total world coal use from 2.5 billion tce in 1977 to about 6–7 billion tce by the year 2000. This is an annual growth rate of 4–4.5 percent. World coal use was in fact increasing at about this rate during the 1950s before declining to an average of 1 percent annually during the 1960–1977 period.

World Coal Import Projections

The teams from countries likely to import coal each calculated their projected import requirements. This was done by estimating the total indigenous production capacity and subtracting it from the total coal demand. A similar exercise was carried out for the regions not directly represented in WOCOL to yield total world coal import requirements.

Although the world's present consumption of coal is large, most coal is consumed in the countries where it is mined. International trade in coal is only about 200 mtce/yr (3 mbdoe) or 8 percent of world coal use. Most coal traded is metallurgical coal for the iron and steel industry. Steam coal represents only about 30 percent of the international trade (less than 1 mbdoe), and most of this is transported over short distances, such as from Poland to the USSR and Western Europe, and from the United States to Canada.

Figure I-4 shows our projections of world coal import requirements for metallurgical and steam coal in Case A and Case B to the end of the century. World coal trade is projected to increase by three to five times to 560–980 mtce in the year 2000. The higher level is equivalent to 13 mbdoe, or nearly half the oil exported in 1979 from OPEC countries.

Figure I-4   World Coal Import Requirements (1960–2000)

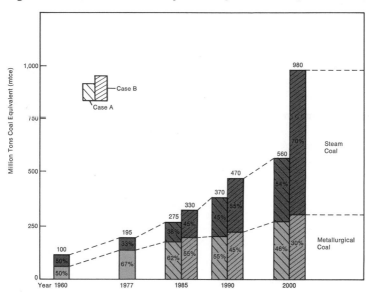

Demand for steam coal imports, which are shown by country and region in Table I-2, are projected to increase even more rapidly by about 5 times in Case A, and about 12 times in Case B. For many nations, the increases are large. Japan, for example, which now imports only 2 mtce/yr of steam coal, is estimated to require 25–50 times as much by the end of the century, and would become the world's largest steam coal importer. Other large coal importers are projected to include France, Italy, and other Western European countries, for example the Federal Republic of Germany and the Netherlands, as well as several newly industrializing countries such as South Korea, Taiwan, and the Philippines in the East and Other Asia region.

The steam coal import requirements for the OECD countries are shown for the Case A variations in Figure I-5. In Case A they grow from 45 mtce in 1977 to 210 mtce in year 2000. The effect of Case A-1 is to increase the requirement to 460 mtce by year 2000. Under the conditions of an oil limitation and nuclear delays in Case A-2, the requirements increase further to 650 mtce by the year 2000,

Table I-2 World Steam Coal Imports by Country and Region (mtce)

Country/Region	1977	1985		2000	
		Case A	Case B	Case A	Case B
Denmark	4.6	10.7	11.1	9.4	20.9
Finland	4.1	3.4	3.4	7.7	12.4
France	14	11	34	26	100
Federal Republic of Germany	3	9	11	20	40
Italy	2.0	10.3	10.9	16.5	45.5
Netherlands	1.5	7.0	7.0	19.9	34.2
Sweden	0.3	2.9	3.2	14.3	23.1
United Kingdom	1	—	—	—	15
Other Western Europe	7	13	13	32	42
OECD Europe	37	67	94	146	333
Canada	6	6	5	8	4
Japan	2	6	7	53	121
Total OECD[a]	45	80	105	210	460
East and Other Asia	—	5	24	60	179
Africa and Latin America	1	3	3	6	10
Centrally Planned Economies	17	20	20	30	30
Total World[a]	60	105	150	300	680

[a] Totals are rounded.

20

Figure I-5   The Effects of Oil Limitations and Nuclear Delays on OECD Steam Coal Import Requirements (1960–2000)

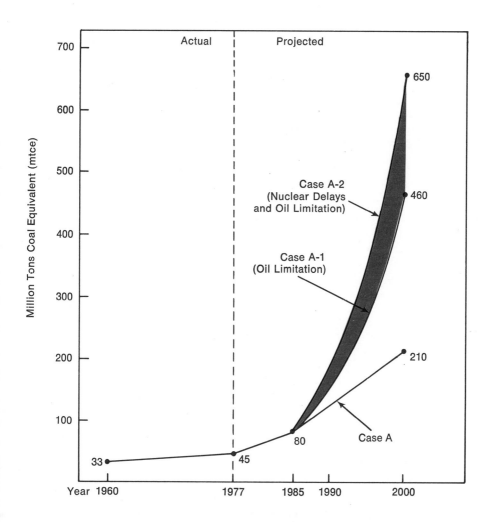

a 15-fold expansion from the 1977 level. The projected expansion of coal imports is similar to the past growth of OECD oil imports which grew 10-fold in the two decades from 1950 to 1970.

World Coal Export Potentials

The feasibility of coal as an international energy source depends on the ability and willingness of coal producing countries to

21

produce, transport, and export large quantities of coal in time to meet the rapidly growing coal import requirements we project. Coal export potentials were estimated for each of the world's major coal producing countries, most of which are represented in WOCOL—Australia, Canada, Federal Republic of Germany, India, Poland, the People's Republic of China, the United Kingdom, and the United States. Current information for Republic of South Africa export potentials by the year 2000 was provided to WOCOL by that government's Department of Environmental Planning and Energy. Our estimates of coal export potentials of the Soviet Union, Colombia, and other possible exporters not participating in the Study were based primarily on data published by the World Energy Conference, the World Bank, and elsewhere.

Table I-3 shows 1977 coal exports and projections of coal export potentials for the year 2000 taken from the WOCOL country

Table I-3 Estimated Coal Export Potentials for Principal Exporters (mtce/yr)

		Year 2000	
			Estimated
		Current	Maximum
Country/Region	1977	Expectations	Potential
United States	49	125–200	350
Australia	38	160	200
Republic of South Africa	12	55–75	100
Canada	12	27–47	67
Poland	39	50	50
Soviet Union	25	50	*
People's Republic of China	3	30	*
Federal Republic of Germany	14	23–25	*
India and Indonesia	1	5	*
Latin America, Africa, others	7	25–50	*
World total	200	550–700	930

* Estimates of maximum export potential were not developed.

and regional studies. The estimates shown under "Current Expectations" were taken from WOCOL Cases A and B and represent what coal exporters now expect to be a plausible range of requirements in the year 2000. The last column, "Estimated Maximum Potential," represents export levels greater than what coal exporters now expect to supply, but which were indicated as feasible for the year 2000 if the demand develops soon enough.

Only Australia (160–200 mtce/yr) and the United States (125–350 mtce/yr) appear to be capable of exporting significantly above 100 mtce by the year 2000. The other significant exporters are projected to be the Republic of South Africa, Canada, Poland, the Soviet Union, and the People's Republic of China.

Some exports are also expected from developing countries such as Colombia, as shown in the estimates for Latin America, Africa, and others. However, these are unlikely to be very large by the year 2000. The difficulty of rapid expansion of coal exports from the developing countries, while meeting growing domestic use, is illustrated by the WOCOL estimates for India and Indonesia of less than 5 mtce total coal exports in the year 2000. This is despite major expansions projected in coal production for domestic use, from 72 mtce in 1977 to 285 mtce in the year 2000 for India, and from negligible levels in 1977 to 20 mtce/yr by the year 2000 in the case of Indonesia.

International Coal Trade Patterns

The largest coal trade flows, as world coal trade expands, are most likely to be from Australia, Canada, Poland, South Africa and the United States to Japan and other East Asian ports and to Western Europe. The implications for both exporters and importers of a coal trade expansion to 600–1,000 mtce/yr are significant, particularly in terms of balance of payments relationships.

Our assessment indicates that coal importing countries could obtain the quantities of coal required from their preferred sources at the lower end of this range with manageable stress on the export capacities of the major coal exporters. At world coal trade levels of 500–600 mtce/yr (Case A), coal importers should be able to diversify their sources of coal.

Diversification opportunities would be greatly reduced however as coal import rates approach 800–1000 mtce/yr, as projected for Cases A-1, A-2 and B. Coal exporters would have to provide much more coal than what they now expect to supply. Moreover, Australia and the United States would have to supply about half of the total world demand for coal imports. The only country where it appears technically and economically feasible to expand coal exports significantly beyond 200 mtce by the year 2000 is the United States. If this were to happen the United States would have to become the world's balancing supplier of steam coal in the 1990s.

Avoiding Delays in Realizing the Required Expansion

Our analysis indicates that, over the next 20 years, world coal production and use is likely to be required to expand by 2.5 to 3 times and provide at least half of all the increase in the world's energy needs. International trade in steam coal will have to grow by ten to fifteen fold.

Given the long lead times involved both for coal using and coal producing projects, the required expansion of coal demand and coal trade will be realized by the year 2000 only if both producers and consumers are willing to make commitments in the early 1980s, even before all the uncertainties about future coal supply and demand are resolved. Unless these commitments are made, there is a real risk that the bulk of the new facilities needed to meet the required acceleration in demand and trade from 1985 onwards will not be available in time. Such delays would limit coal's contribution to the levels projected in Case A which, while large, would be inadequate to compensate for the projected leveling of world oil supplies and possible delays in nuclear expansion programs, even for the lowest levels of energy demand growth that we have investigated.

Coal Markets and Prices

Because of low demand relative to production capacity, a widely distributed resource base, and high transport costs relative to price at the mine, steam coal markets have tended to be geographically limited. Metallurgical coal markets, on the other hand, have been more worldwide than steam coal markets, because metallurgical coal has a smaller, less widely distributed resource base, almost no substitutes in use, and has traditionally received higher prices than steam coal.

Even before the doubling of oil prices in 1979, however, steam coal was competitive in many national and regional energy markets. Since then it has become increasingly more economically attractive in international markets, especially in the electric power market. This is despite the large maritime transport costs from suppliers such as Australia, Canada, the United States, and South Africa to consumers in Western Europe and Asia, as well as the high costs of environmental control in some of the consuming countries. Further increases in the price of oil should continue to improve the competitive position of steam coal and improve its market penetration in both geographical and end-use markets.

In contrast to an increasingly constrained world oil supply, coal has a very large resource base and considerable scope for expanded production. Given sufficient lead time, it should be possible to continue to increase the supply of coal to meet additional demand at costs that remain competitive. Moreover, the growing number of international coal suppliers, each with different interests, from all world regions including OECD, CPE, and the developing countries, makes the formation of an international coal cartel unlikely.

The growing future demand for steam coal projected by WOCOL will be accompanied by a large expansion in world coal trade. This worldwide market in steam coal will involve many suppliers and will include joint ventures between consumers and producers. Steam coal will be purchased both under long-term contracts and on spot markets. Other arrangements will include captive mines and contract mining. This international steam coal market may tend to acquire some of the characteristics of an international commodity market.

If steam coal is to make the necessary contribution to world energy needs in the coming decades, it will have to remain competitive with other sources of energy, including unconventional oil and gas. In the steam coal market, this competitiveness will be determined by coal's price per unit of obtainable heat, after adjusting for handling and utilization, and other quality characteristics.

Over the next two decades coal will primarily go into the heat and steam-raising markets. It is here that interfuel competition is strongest, although substitution is constrained by the user facilities and infrastructure in existence as a result of previous energy investment decisions. The outcome of the ongoing nuclear debate and the success, or otherwise, of the energy conservation programs now being proposed and implemented will also affect the demand for coal and the prices it can command in the energy market.

Over the long term, the real cost of coal is likely to rise for a number of reasons. The development of new and more costly mining facilities to replace or expand existing capacity; meeting new health and safety requirements; increasing expenditures on environmental protection and land reclamation; and meeting the rising costs of labor will all contribute to future cost increases. On the other hand, increasing productivity, greater mechanization, larger ships, and other economies of scale should help to moderate such increases.

The short-term behavior of prices can, of course, differ substantially from long-term trends. If production capacity exceeds demand, as it did in 1979, prices will tend to move toward the level of the variable cost of marginal mines rather than being determined by the price of alternatives. On the other hand, if the quantity of coal demanded at a particular price exceeds the supply, because of short-term supply and demand inelasticities, spot prices will tend to rise as producers and transporters attempt to capture the increased short-term revenue. However, such price increases would also provide economic signals to some coal suppliers to increase coal production, and thus set forces in motion to eliminate continuation of the short-term advantage. Moreover, a large proportion of the future trade in coal will not be priced on a spot basis. It will be the subject of long-term contracts with greater price stability, and with prices more reflective of the cost of new as well as existing production and transport facilities.

These long-term production and transport costs will set the lower boundary to future prices. The upper level will be determined by the price of the best available alternative. However, in the long run, because of its abundance, and provided a free and competitive international market is maintained, there is little reason to expect that steam coal prices will be directly coupled to world oil prices. This will be particularly so as oil is increasingly removed from the heating market for use in transport and as a specialized petrochemical feedstock.

Environment, Health, and Safety

General Issues

Public acceptance of a large expansion in coal use will be influenced by evidence that coal can be mined, moved, and used in ways that permit adherence to high standards of environmental quality. In this Study we have examined the environmental standards currently in effect or under consideration in WOCOL countries; we have considered the environmental control technologies that are available to meet these standards, and we have estimated their costs. Our findings are summarized here and presented in detail in Chapter 4.

The principal areas of public concern include land reclamation after surface mining; subsidence from underground mining; acid drainage from the refuse from coal mines and coal preparation plants; emissions from combustion such as SO_2, NO_x, and particulates; safe disposal of ashes; and the possible effects of CO_2 on climate. The applicability and priority of these concerns vary from country to country depending on a number of circumstances. Except for the CO_2 question, however, technology is available to meet these concerns and to comply with the most stringent of the current environmental standards in each WOCOL country at costs that leave coal competitive with oil at mid-1979 prices in most areas. There is no practical method of controlling CO_2 emission from the combustion of fossil fuels and from other sources, and further research is needed on the possible effects of increased CO_2 emissions on global climate. Control of long-range transport of gaseous and particulate emissions may also require new forms of international cooperation.

Difficult choices are required to achieve an appropriate balance between the degree of control of emissions and the resource costs necessary to achieve that control; these, in turn, must be related to the benefits arising from the use of coal. But each increment of further control increases the cost, and as total control is approached the costs become very large. Each country makes its own judgments based upon its own circumstances; we have assumed that national environmental standards will continue to reflect these judgments.

Beyond this, however, it is clear that further research is needed to lower the costs of installing and operating pollution control devices and to improve their effectiveness. Moreover, additional research is required to identify those effluents causing the greatest risks to assure that adequate, but not excessive, control is exercised. Nevertheless, within what can be reasonably anticipated to be applied as regulation, coal can be expected to remain competitive with other fuels in most locations.

Mining Issues

It is now possible with current technology and at a reasonable cost to restore most surface-mined lands to a condition equal to or better than their original condition. Reclamation is generally easier for flat areas. On steep mountain slopes, in arid regions, or in certain ecological regions, reclamation is more difficult. Mining may not be permitted in some such areas under the new laws in effect in several major coal-producing countries. Land areas critical for other purposes such as agriculture may also be excluded from mining. Australia, Canada, the Federal Republic of Germany, the United States, and the United Kingdom all have comprehensive legislation for the control and reclamation of surface mined lands. Even with the possible limitation of mining in some areas, because of the cost of reclamation or because of regulations limiting their accessibility, there are still more than sufficient coal resources in mineable terrain to meet the coal demands projected by this study.

Underground mining can cause land subsidence, i.e. sinking into the area that has been mined. Where the room-and-pillar method of mining has been used, the problem of subsidence is usually not great. However, the room-and-pillar method leaves as much as half the coal in the ground. Another approach is to allow the surface of the land to subside after the removal of the underground coal, and to provide for quickly and fairly carrying out repairs or giving compensation for any damage that may occur. This approach can be used only under certain geological conditions and usually when the longwall method of mining is used. It is currently employed in the United Kingdom and other European countries.

Large quantities of solid waste are produced from surface and underground mining, as well as from coal preparation plants. This must be disposed of in surface piles, which can be landscaped; as landfill; by returning to the mine; or by use as a construction material. In areas where the waste material contains contaminants, such as high levels of sulfur, prevention of leaching requires careful control of water flows near the storage area, or may even require special ponding arrangements.

Occupational health and safety are important concerns in coal mining. The major occupational health problems associated with underground mining has been the lung (pneumoconiosis or black lung) disease caused by breathing coal dust. The greatest safety hazards in underground mining have been from gas explosion and flooding. Reduction of dust and gas levels by much improved ventilation and filtration systems; dust suppression by water-spraying, or laying powdered limestone; continuous monitoring of air quality and the application of strict work rules and practices have done much to reduce the risk of lung disease and to improve safety by reducing the risk of explosion. Moreover, increasing mechanization and better equipment have reduced the exposure of the work force per unit of coal produced. In mines where the best current practices are observed, the accident and illness rates are now comparable with construction work and many sectors of heavy industry. As coal production expands, extending the application of these health and safety practices and continuing worker training should ensure acceptable occupational health and safety conditions.

Surface and underground coal mining can both be managed with acceptable environmental, health, and safety consequences at reasonable costs. This does not, however, mean that mining procedures in some countries may not need to change. In the United States, for example, producers are now allowed little flexibility in meeting current standards. As experience is accumulated it may be possible to meet the required standards at lower costs through different techniques, and such flexibility should be encouraged.

Coal Transportation and Storage Issues

Coal can be transported by conveyor or truck for short distances, or by train, barge, ship, or pipeline for long distances. The principal environmental disturbances are generally dust, noise, congestion, and runoff of contaminated water. In contrast with oil, the risk of environmental effects from spills are very small, but there are risks from accidents with trains or trucks.

Dust can be controlled by spraying, compacting, or covering. Water runoff can be controlled by careful design of coal storage facilities and treatment of the water where the coal contains soluble or leachable contaminants. Other environmental effects can be controlled by the appropriate design of storage facilities.

Coal Use Issues

Virtually all coal use involves combustion of at least part of the coal with a release of a number of solid and gaseous substances.

Most countries now have air quality standards and regulations that limit the rate of allowable emissions from sources such as coal-fired electric power plants. The major emissions that are regulated include sulfur dioxide (SO_2), particulate matter (total suspended particulates or TSP), and nitrogen oxides (NO_x). The problem of reduced visibility in cities such as London has been greatly diminished by the use of smokeless fuels or the change to appliances using gas, oil, or electricity by residential and commercial consumers.

Technologies for controlling the solid and gaseous emissions from coal combustion already exist, and improved technologies, either to lower costs or to reduce emissions even further, are being developed. Greatly reducing such emissions results in high costs for emission control. Because there are substantial areas of disagreement among experts as to the effects of these emissions, it is not surprising that national policies differ widely on emission control goals and strategies. Cleaning up some of the emissions, especially sulfur, may create new waste disposal problems such as limestone sludge from flue gas desulfurization.

The extent of the environmental impact from coal-based synthetic fuels production is not clear. Synthetic fuel processes are complex. There are currently very few data available on the characteristics of emissions from the various possible conversion processes because no commercial scale plants have been built except in South Africa. The major environmental problems will be controlling the production and release of potential carcinogens (primarily, complex organic compounds) during the coal conversion and possible toxic materials in the waste. Because the basic costs of coal conversion will almost certainly be high, the additional costs of controlling emissions will probably be an acceptable fraction of the total cost given the market values and clean nature of liquid and gaseous fuels.

There are some issues on which joint action among nations appears necessary. For example, acid rain resulting from the long-range transport of emissions including those from coal burning is acute in some regions and may require early actions by nations in such a region, for example Western Europe. New mechanisms for international cooperation may be needed. OECD studies of transnational pollution are leading to some agreement on sources of sulfur in the atmosphere (domestic or imported) and on procedures for consultation.

Carbon Dioxide

All hydrocarbons, wood, petroleum, natural gas, vegetable matter, and coal produce carbon dioxide when they are burned. CO_2 is also produced in the respiration of living creatures. Any problems arising from CO_2 emissions are therefore not unique to coal. However because coal is mainly carbon, it releases 25 percent more CO_2 than oil and 75 percent more than natural gas per unit of heat produced.

There is no practical way of controlling the emission of CO_2 from fossil fuel combustion. It is generally agreed that the CO_2 concentration in the atmosphere has recently been rising at about 0.4 percent per year and that a continuation of that rise could have a warming effect on the atmosphere. But there is still great uncertainty about most other aspects of the global carbon cycle, about other factors affecting climate, and about the effects a global warming might have on man's activities in different areas of the globe. The most authoritative recent statement on the CO_2 question was issued by the World Climate Conference held in Geneva in February 1979. The statement said:

> The causes of climatic variations are becoming better understood, but uncertainty exists about many of them and their relative importance. Nevertheless, we can say with some confidence that the burning of fossil fuels, deforestation, and changes in land use have increased the amount of carbon dioxide in the atmosphere by about 15% during the last century and it is at present increasing at about 0.4% per year. . . . It is possible that some effects on a regional and global scale may be detectable before the end of this century and become significant before the middle of the next century. This time scale is similar to that required to redirect, if necessary, the operation of many aspects of the world economy, including agriculture and the production of energy.

It is our view that there is time to conduct the necessary further research on the possible effects of CO_2 from fossil fuel combustion on climate. There is evidence of increasing support for such research, and extensive international collaboration through such bodies as the World Meteorological Organization of the United Nations. We also believe that the present state of knowledge about CO_2

effects on climate does not justify action to limit or reduce the global use of fossil fuels or delay the expansion of coal use even if a mechanism for such concerted actions by all nations were available.

Attitudes and Constraints

Attitudes differ among countries toward such things as exposure to health risks or the importance of avoiding reductions in visibility. Environmental impacts also vary because of differences in topography, climate, or population density. The importance attached to particular environmental risks will also be related to the wealth of a country, its attitude toward industrial growth, and the availability of alternative energy supplies. For these and other reasons, nations and regions will differ on the kind and extent of environmental control measures they will require as coal use increases.

Coal production, transport, storage, and use require the use of land and frequently also require the use of water. Because both land and water resources have competing uses, and because in areas of dense population this competition is acute, there need to be ways of allocating the rights to use land or water. Each nation has developed such patterns of allocation by custom, marketplace, regulation, administrative fiat, or legal processes. In some countries and localities, increasing difficulties are leading to development of comprehensive planning for allocation of land and water among competing needs.

Research and Development

Research leading to a better understanding of the environmental effects of coal production and use must be continued; so also must the development of improved methods of environmental control. These will be invaluable if emission standards become more stringent than they are today or if issues which we have judged not serious require additional control.

Major constraints to a rapid increase in coal use are the complex and constantly changing environmental protection procedures that have evolved in some countries. These make for lengthy and costly delays in siting and building coal-producing, transport, and using facilities. Investments will be delayed or will not be made at all, unless a consensus is developed about the necessity for coal, and unless some of the approval procedures become more simplified and certain.

Values and standards will prevent some coal from being mined and used in some areas. We are, however, convinced that coal can be produced, moved, and used in ways which meet the environmental standards of each particular country concerned. The technology to do so is, in most cases, already available. Further technological advances can greatly widen the scope for the environmentally acceptable use of coal.

Coal Resources, Reserves, and Production

World Coal Resources and Reserves

There are vast resources of coal in the world, far in excess of those of any other fossil fuel. This resource base is sufficient to support greatly expanded worldwide use of coal well into the future.

Coal resources vary substantially in quality, reflecting the complex chemical structure of coal. The energy content or calorific value of the coal and the sulfur, ash, and moisture content are some of the key elements of quality. Systems of coal resources classification have been developed to deal with these complexities; however, the use of these systems varies from country to country.

The most commonly used classification standards are those established by the World Energy Conference (WEC), which has defined *geological resources* of coal as a measure of the amount of coal in place, and *technically and economically recoverable reserves* as a measure of the quantities that can be economically mined with current mining technology and at current energy prices.[6] The WEC also subdivided coal into two major calorific categories: *hard coal*, defined as any coal with a heating value above 5,700 kilocalories per kilogram (equivalent to 10,250 Btu per pound) on a moisture and ash-free basis; and *brown coal,* any coal with a lower heating value. Bituminous and anthracite coal fall into the category of hard coal, whereas lignite and subbituminous coal are considered by the WEC as brown coals. However, some countries, for example the United States and Canada, use the term "brown coal" generally to refer only

---

6. Chapter 5 provides a complete description of the WEC standards, which are based on specific definitions of both the depth and thickness of coal seams, and describes in more detail other aspects of coal reserves and production.

to low-calorific lignite and not to the extensive resources of subbituminous coal that they possess.

Figure I-6 illustrates the quantities of geological resources and
technically and economically recoverable reserves of hard coal and

Figure I-6
Geological Coal Resources of the World (in 10⁹ tce)

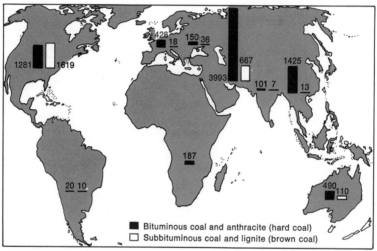

Source: 1978 WEC figures updated by WOCOL country teams. Non-WOCOL countries
updated by Bergbau-Forschung.

Economically Recoverable Coal Reserves of the World (10⁹ tce)

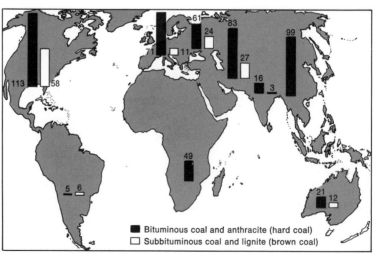

Source: 1978 WEC figures updated by WOCOL country teams. Non-WOCOL countries
updated by Bergbau-Forschung.

brown coal by regions on a tons of coal equivalent (tce) basis. The estimates are based on the 1978 WEC report[7] as updated by some WOCOL country teams, such as Australia.

Coal resources and reserves are geographically widely distributed with more than 80 countries reporting coal deposits. However, 10 countries account for about 98 percent of the currently estimated world resources and 90 percent of the reserves. Moreover, 4 countries, the Soviet Union, the United States, the People's Republic of China, and Australia, possess almost 90 percent of the total world coal resources and 60 percent of the reserves.

The magnitude of the world's coal resources and reserves is difficult to comprehend. Technically and economically recoverable coal reserves currently amount to about 660 billion tce or approximately 250 times the 1977 world production. Moreover, estimates of reserves increase as exploration reveals further quantities of extractable coal and as economics and technology change. As a result of increased exploration, stimulated by the oil price increase since 1973, estimates of the world's technically and economically recoverable reserves have increased by about 185 billion tce. This increase is equivalent to about 70 years production at the 1977 rate of production.

World Coal Production

World coal production in 1977 was about 2,500 mtce—equivalent on an energy basis to 33 mbdoe or more than the total OPEC oil production. Table I-4 shows 1977 world coal production as well as projections of production for WOCOL Case B for the year 2000. The WEC, the World Bank, and other sources were consulted for the coal production estimates for countries and regions outside WOCOL.

Three countries—the United States, the Soviet Union, and the People's Republic of China—were responsible for nearly 60 percent of the total coal produced in the world in 1977. The next six largest producers, Poland, the Federal Republic of Germany, the United Kingdom, Australia, the Republic of South Africa, and India, accounted for a further 25 percent. In 1977 the United States accounted for 50 percent of the total production outside the centrally planned

---

7. "Coal Resources," World Energy Conference (1978).

Table I-4   Coal Production for Major Coal-Producing Countries Actual and Projected—1977 and 2000 (mtce/yr)

Country	1977	2000[b]
Australia[a]	76	326
Canada[a]	23	159
People's Republic of China[a]	373	1450
Federal Republic of Germany[a]	120	150
India[a]	72	285
Poland[a]	167	313
Republic of South Africa	73	228
United Kingdom[a]	108	162
United States[a]	560	1883
Soviet Union	510	1100
Other countries	368	724
Total world	2450	6780

[a] WOCOL member
[b] Based on WOCOL Case B
Source: WOCOL country reports, plus WEC, World Bank, and others for countries outside WOCOL.

economy countries. Production in the developing countries is similarly concentrated, with India being by far the largest coal producer and consumer.

Even with the significant expansion in world coal production projected by WOCOL, the cumulative production during 1977-2000 would use up just 16 percent of the world's present technically and economically recoverable coal reserves.

Coal's Role in Developing Countries

Because oil was inexpensive, convenient to transport and use, and readily available until the early 1970s, exploration for coal in the developing countries has been much less widespread and less intensive than exploration for oil and natural gas. Much of the present resources of coal are located in the northern temperate zone. Although the southern hemisphere is less favorable for coal deposits from a geological viewpoint (i.e. less extensive large sedimentary basins), there are large coal resources in Australia and the Republic of South Africa, and there is some optimism that expanded exploration in the southern hemisphere and in the less developed regions of the northern hemisphere will result in the discovery of significant new coal reserves. For example, recent exploration in the southern part of Africa, particularly Botswana and Tanzania, as well as in Indo-

36

nesia, is yielding favorable results. The world's coal resources and reserves could be significantly larger and more widely distributed geographically than was previously thought.

The World Bank in its recent publication, *Coal Development Potential and Prospects*, notes that coal production in the developing countries accounted for only about 5 percent of the total 1977 world coal production. About 50 developing countries have known coal resources, and about 30 of these are currently producing coal. A large expansion in coal production and use was projected in this report for the developing countries. In addition to meeting an increasing share of their domestic energy needs, some of these countries, notably Colombia but also Indonesia and Botswana for example, may have significant potential for exporting coal in the future.

Rapid coal development in the developing countries will require both increased production and increased domestic use of coal. Many of the developing countries have neither the financial resources nor the technical and managerial know-how to launch major coal development programs on their own. Thus international, regional, and bilateral agencies, as well as private mining companies, are likely to be required to play a major role in supporting the developing countries in an analysis of their coal potential, in an assessment of the role that coal could play in their total energy supply balances, and in providing, where appropriate, financial and technical support for the implementation of coal projects.

Coal-Mining Methods

The choice of whether coal will be mined by surface or underground methods is determined almost entirely by the geological characteristics of the deposit. In general, only deposits that are at shallow depths can be surface mined, but sometimes when the coal occurs in multiple seams, it becomes possible to surface mine at considerable depths.

Surface mining is usually less costly than underground mining and permits the recovery of a higher proportion of the coal in place. Surface mining is also less labor intensive than is underground mining. The majority of the world's coal deposits are, however, accessible only by underground mining. Even though surface mining will expand significantly over the coming decades, particularly in some countries, for example the United States, it is probable that by the end of the century and thereafter the greater part of the world's coal production will come from underground mines.

Underground mining is usually carried out by either the room-and-pillar method or the longwall method. In the room-and-pillar method, which is the most common in the United States, the coal is extracted leaving behind pillars of coal that support the roof and that considerably lessen, if not eliminate, the possibility of ground subsidence above the mine. However, this method leaves about 40–50 percent of the coal in the mine.

The longwall method extracts the coal across the whole cutting face of a seam. Powered supports are used to hold up the roof temporarily while the coal is cut and removed. The supports are then removed and the roof is allowed to collapse. The longwall method allows a higher proportion of the coal to be removed than does the room-and-pillar method and, because it tends to be more heavily automated, it has a generally higher productivity. It is the most commonly used method in European mines.

Future Prospects

Coal reserves are sufficient to support a large expansion of coal use well into the next century and beyond. The technology for coal mining is highly developed and steadily improving. Although a large fraction of currently known coal reserves are located in relatively few countries, coal resources exist in many countries throughout the world. There are good prospects that these coal resources can be developed to support both domestic use and additional exports in many more countries in the future.

Ports and Maritime Transport

The worldwide expansion of steam coal production and trade projected by the World Coal Study will require substantial infrastructure development, including a major expansion of domestic and international transportation facilities. Moreover, transportation costs are usually an important element in the total delivered cost of coal.

Figure I-7 illustrates the transport links in a chain from a coal mine to a final market for coal.[8] The simplest chain is from a

---

8. Chapter 6 contains a detailed discussion of maritime transport and port issues. The country reports in Volume II include descriptions of the specific inland transport and port requirements within the WOCOL countries.

Figure 1-7   Generalized Domestic and International Coal Chains and Markets

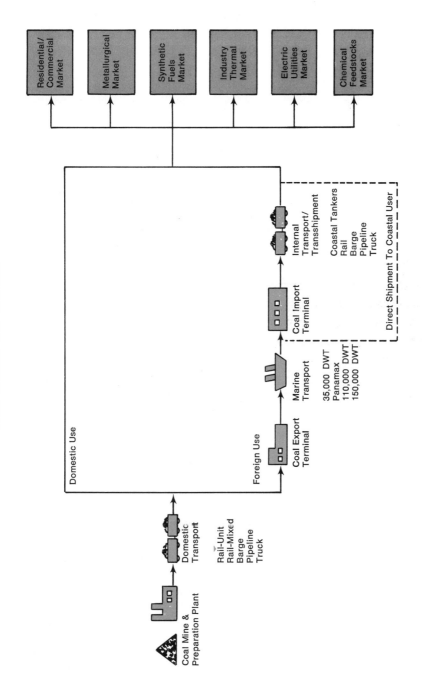

mine to a nearby domestic market that may in some cases be located next to the mine site. Coal for domestic use moves by conveyors, railroads, unit trains, slurry pipelines, barges, and trucks. Coal in world trade moves through export ports into various types of ships to receiving ports and then into local distribution systems.

Inland Transport

A major expansion of railway and barge transport systems, which now carry the bulk of the coal transported inland over long distances, will be required. This expansion will be particularly large in countries where long distances are involved, such as Canada and the United States.

In some countries an attractive alternative to coal trains is the use of pipelines to transport coal mixed as a slurry in water. These pipelines are expected to be able to deliver coal at about the same, or lower, cost than railroads under high-volume, long-distance load conditions. Because they can be buried and are virtually noiseless, pipelines may be the only acceptable way of expanding the transport of coal in densely populated areas where the frequent passage of mile-long unit trains might prove unacceptable. However, the development of slurry pipeline systems currently faces difficulties in some countries in obtaining rights-of-way. Moreover, they need a mix of a ton of water for each ton of coal and so may be restricted to regions with adequate water near the coal source. Returning the water by pipeline or bringing water from a distant external source can, however, solve this problem, although with some increases in the transportation cost.

Tests are now also being conducted on the use of coal-oil slurries. If successful they may provide the quickest and simplest way of using coal in power plants designed for oil. Moreover, these coal-oil slurries could in many cases be moved in existing oil transportation and distribution systems and therefore could somewhat reduce the need for new coal transportation facilities.

Lead times are long for all forms of transport systems, and planning and approval procedures in some countries can take much longer than the construction of facilities. The choice of new railway routes is for example restricted by the need to avoid steep gradients, and residents along the routes may have objections to noise, dust, or other impacts associated with greatly increased traffic. Moreover, upgrading the capacity of existing rail lines may place burdens on the financial resources of some railroads.

Maritime Transport

Maritime transportation costs are and will continue to be a very significant part of the total delivered cost of imported coal. The coal trade routes are generally long, mainly from North America, Australia, or South Africa to Western Europe or East Asia. Transportation costs can therefore be the deciding factor in establishing the balance of competition among these distant sources of coal.

Coal is carried either in dry bulk vessels which are also capable of carrying other commodities such as iron ore, grain, bauxite, and phosphate rock or in so-called combination-carriers that can carry oil as well as dry bulk cargoes. Of the dry bulk commodities, iron ore, grain, and coal, in that order, are by far the most significant. The relative significance of coal as a commodity in international trade will increase significantly by the year 2000 under the WOCOL projections.

A major building program will be required to provide the new ships involved in realizing the projected expansion of world coal trade. The shipbuilding requirements are, however, well within the capacity of the industry. In fact, the building program—averaging 50 coal ships or 5 million DWT per year for 20 years—would be of great economic benefit to the world shipbuilding industry.

In general, the larger the ship the lower will be the unit costs and hence the lower the freight rate. The present maximum size of ships operating in the dry bulk trade is about 150,000 DWT. During the 1980s the major part of coal in seaborne international trade is expected to be carried in vessels of about 100,000–125,000 DWT. In the 1990s, ships as large as 250,000 DWT may come into use. The increasing ship sizes will have to be accompanied by development of some coal ports that can provide the draft depths and berths to accommodate large ships. Many coal ports, however, have neither the draft nor the other necessary dimensions to receive large bulk carriers. In some cases these larger ships will off-load part of their shipment at deep-water ports and the balance at ports with less draft. Some ports will develop into coal storage and distribution centers. The mix of ships will therefore continue to include smaller ships such as Panamax-size vessels (65–75,000 DWT) that can pass through the

Panama and Suez canals, self-loaders such as those of the Great Lakes in the United States, barges for river transport, and small coastal ships.

Coal was the original bunker fuel for the first mechanically propelled vessels, but it was displaced with the advent of cheap oil. Present and expected future constraints on the availability and the price of oil are causing serious reexaminations of the potential use of coal as bunker fuel for ocean-going vessels. Coal-fired ships require a greater capital cost but appear increasingly attractive economically, because of expected fuel cost savings, and by the year 2000 a substantial fraction of the world merchant fleet may be coal-fired. Initially this would be expected for coal-carrying ships; wider application would require coaling stations along trade routes such as existed in the past. Some Australian companies have completed feasibility studies and are inviting shipbuilders to bid for coal-fired bulk carriers, for transporting coal to Asian ports and minerals around the Australian coast. The use of coal-fired ships on longer routes to Europe will require establishment of bunkering facilities along these routes.

Export Ports

Most coal export and import facilities were designed to handle metallurgical coals that require extensive blending and storage facilities at export ports. The requirements for steam coal exports are simpler.

Terminals—export or import at coal ports—will have to accommodate the 150,000 DWT or even larger vessels that will be used on some routes. Facilities for unloading unit trains and perhaps pipeline terminals as well, will be needed. Suitable berths, draft depths, and storage areas as well as rapid ship loading and unloading equipment are necessary parts of a modern coal terminal.

A significant expansion in coal export facilities will be required to meet the projected increase in demand for exports. These requirements have been carefully investigated in the country reports prepared by coal-exporting countries in WOCOL. For example, increasing Australian coal exports would require upgrading and expansion of existing ports as well as the construction of several new ports. The diversity and geographic distribution of projected steam coal export markets from the United States would necessitate the establishment of new coal export facilities on the Gulf and West coasts as well as new and expanded facilities on the East Coast, where the bulk of coal export facilities now exist. All WOCOL exporter countries concluded that coal export facilities could be expanded with only moderate capital investments.

However, the regulatory and institutional processes for obtaining approvals for port expansion projects are complex in many countries. Frequently the approval procedure may involve several independent governmental organizations with local or national jurisdictions. These regulatory and institutional considerations are likely to be the most significant potential constraint to the timely development of the required port capacity. In some countries it may be necessary for governments to facilitate the process for siting and building new coal export facilities.

Import Ports

Coal can be imported directly to the user, for example, a coastal power plant, or to a terminal that receives large volumes of coal for many customers.

Around the coasts of Europe an established pattern of coastal ship movement exists which is capable of expansion to handle much larger coal tonnages. Inland, the capacity of river, canal, and rail systems may already be adequate. However, existing terminals capable of receiving coal from large carriers, storing it and transshipping to its final destination may be insufficient. A number of sites exist or are being studied for possible development in Denmark, Sweden, the Netherlands, France, and Italy.

Capacity already exists in Japan to handle the projected increases in imports of metallurgical coal. Some steam coal imports may be handled in designated ports associated with specific coastal power stations. The construction of large "coal centers" or transshipping terminals is also under study.

The role of smaller ports may also be important in some areas. Where ports are limited in draft only and otherwise capable of accepting large carriers, they may be served by large ships that have unloaded part of their cargo at another port that is not subject to such draft limitation.

Overall, it would appear that as is the case for coal export terminals, the timing of import terminal expansion is likely to be a more critical factor in handling coal imports than the physical feasibility or costs of the expansion.

Transport Systems

Perhaps the most important aspect of all in looking at future transport of coal is the need to consider transport within the context of a coal chain. Coal transport systems—inland transportation, export ports, maritime transport, and local distribution—are the essential links between producers and consumers although frequently independent of both. Unless transport development is integrated with the planning for mines, power stations, and other coal facilities, delays will occur and coal trade will fail to expand at the rate required.

Technologies for Using Coal

Established Technologies

Coal is already used extensively and the technology for expanding its use is well tried and readily available. This applies particularly to the direct combustion of coal to produce heat and steam in electric power generation and in industry. Engineers have been designing and using coal-fired boilers for the past 200 years and the technology has reached an advanced state. Much progress has also been made over the past few decades in improving the efficiency of electricity generation. The best modern coal power stations can now achieve an efficiency of up to 40 percent, close to the maximum possible with conventional materials.

Substantial developmental activity is also in progress to further improve control of particulate and gaseous emissions from combustion, in order to meet increasingly strict environmental regulations. Considerable progress has been made in this area in recent years. Reductions of 99.8% in particulate emissions are for example achievable using either electrostatic precipitators or bag houses. Most of the emphasis at present is on reduction of sulfur oxide emissions. Wet scrubbers in which the sulfur is absorbed are in commercial use, and regenerative scrubbing systems, which avoid the sludge disposal problems of the wet systems, are at an advanced stage of development.

There have also been advances in the technology for smaller scale uses of coal in residential/commercial markets. The technology is now available for burning bituminous coal in clean air zones in some countries. Such methods including automatic underfeed stokers are suitable for district heating plants, apartment houses, hospitals and other small or medium scale applications. Systems for further improving the convenience of coal use in the domestic sector are also under development.

Developing Technologies for Direct Use of Coal

Many countries are active in developing new technologies, or improving existing ones, to widen the markets for coal combustion. These efforts are directed primarily to increasing the efficiency and flexibility of coal combustion, extending the range of coals that may be used in individual furnaces, and decreasing the environmental impacts associated with coal combustion.

Fluidized bed combustion is one of the most promising of the new technologies for coal combustion. Fluidized bed combustion at atmospheric pressure has been developed to a stage where manufacturers are prepared to give performance guarantees on industrial scale boilers. It is also anticipated that this technology will be developed so that it can be utilized for centralized power generating stations (in multiples of 200-300 MWe units) within the next ten years. Pressurized fluid bed combustion units are at a substantially earlier stage of development than the atmospheric pressure units.

The fluidized bed combustion method offers a number of important advantages over conventional boilers. The absorption of sulfur oxides within the bed minimizes the problem of removing these from the flue gas. Furthermore, the low combustion temperature reduces the formation of nitrogen oxides substantially. It is also possible to use lower-quality coals which yield very high quantities of ash. Finally, because fluidized bed units are more efficient than conventional boilers, they can be smaller than conventional units of the same capacity.

Progress is also being made in the development of technology for dispersing pulverized coal in oil, called coal-oil mixtures (COM), to the extent of 20–50 percent coal, and for using COM as a fuel in furnaces and boilers designed for oil firing. There is considerable interest particularly in the United States and Japan in the prospects for using COM fuels as a way of accelerating the substitution of coal for oil in existing power plants. For new boiler capacity, the most economic approach will continue to be designing directly for coal combustion.

Developing Technologies for Conversion of Coal to Gaseous and Liquid Fuels

The development of technologically reliable and economically viable methods for the large-scale production of gaseous and liquid fuels from coal could greatly widen the scope of markets for the use of coal in the 1990s and beyond.[9] Interest in technologies to convert coal to oil and gas has greatly intensified recently as prospects for the future availability and price of conventional petroleum sources, especially oil, have worsened.

Because the ratio of hydrogen to carbon in coals is usually less than in petroleum-based fuel, the conversion of coals to gaseous or liquid fuels requires complex chemical changes and the introduction of additional hydrogen, as well as the parallel elimination of oxygen, sulfur, and nitrogen.

Gasification of coal involves the reaction of the coal with steam and requires a heat source, usually provided by having oxygen present. Several grades of gas, in terms of their calorific or heating value, are produced from coal. There is also substantial interest in upgrading such gas to produce methane, i.e., substitute natural gas (SNG), which offers the possibility of supplementing natural gas supplies in existing pipeline systems and for existing appliances.

There are many coal gasification processes under development. The variety reflects the great heterogeneity and variability of coal as a raw material and the complex chemistry involved in conversion. The Lurgi process is probably the best known. It was developed to the stage of large-scale commercial operation using subbituminous coals and lignites before World War II, and further developed in the early 1950s to permit use of noncaking bituminous coals. It is now used in the Republic of South Africa to produce medium-calorific value gas for use as a fuel as well as an intermediate synthesis gas for the manufacture of motor fuels and various chemicals using Fischer-Tropsch synthesis. Several Lurgi plants are also in operation in Eastern European countries.

---

9. A more detailed treatment of gasification and liquefaction technologies is provided in Chapter 7.

The Winkler fluidized bed and the Koppers-Totzek (KT) gasification processes are also in industrial use at present to produce an intermediate synthesis gas in fertilizer and ammonia plants respectively. A number of new gasification processes are also under development in several countries, with the common objective of allowing gasification to proceed at elevated pressures.

There are three general approaches to the production of liquid fuels from coal: (1) pyrolysis, (2) hydrogenation, and (3) gasification followed by conversion of the synthesis gas to liquids using Fischer-Tropsch or other technologies. All coal liquefaction processes require some gasification of coal as well to produce the additional hydrogen required. Liquid fuels were produced from coal and coal tar in Germany during the Second World War using hydrogenation and the Fischer Tropsch method. Liquefaction plants were also commissioned in other countries. However, the increasing availability of cheap oil and natural gas after World War II removed much of the incentive for any further work on liquefaction and brought about the shutdown of existing coal liquefaction plants.

The only coal liquefaction plant in commercial use today is located in the Republic of South Africa and has been operating since 1955. This plant (SASOL) which uses the gasification/Fischer-Tropsch approach, has an annual production capacity of 240,000 tons/year (4800 bdoe) of liquid fuels. A second plant is now being commissioned that will increase production significantly and a third plant has been announced.

Several other coal liquefaction processes are under active development at present. These aim to produce a range of products from clean liquid power station fuels through to a synthetic crude oil capable of being refined to a variety of liquid fuels. There is now significant government support for process development in several countries, as reflected in the WOCOL projections.

Lead Times and Opportunities

It is important to distinguish between the time needed to build a plant using commercially demonstrated technology and that needed for the development of a new technology to the point at which it can be considered ready for commercial use. Experience shows that a period of about 30 years is necessary for the full scale development of any major new processing technology from the initial concept. It is therefore impossible to make precise estimates of the timing of future commercial use for new technologies still under development and which still face technical problems that must be resolved.

Even conventional power stations, which are now usually built in units of 400 to 750 MW in the industrialized countries, require a period of 6 to 7 years for completion, about 4 years of which is needed for construction. In some cases, the total timespan from project initiation to commissioning of the plant can be up to 10 years, the extra time arising from administrative procedures, including the need for approvals from regulatory bodies that often involve public hearings and delays. For the established gasification technologies and the Fischer-Tropsch liquefaction process, construction times are comparable with power stations.

This question of lead times is crucial. Substantial expansion in coal use within this century must be based largely on currently available technology, that is, conventional power-stations, industrial boilers and furnaces, and established domestic and commercial uses. If coal-derived liquids and gases are to make any significant energy contribution before the year 2000, the aggressive development programs currently planned for pilot plants and prototype large-scale facilities must be followed up quickly by large commitments for commercial facilities.

The technology upon which to base a rapid expansion of coal use is well established and steadily improving. Moreover, the continuing technological advances in the newer combustion technologies, such as fluidized bed combustion and coal-oil mixtures, offer an opportunity for widening the application of coal combustion or improving the environmental characteristics of its use. New gasification and liquefaction processes are undoubtedly going to be essential elements in the future energy strategies of many nations in the 1990s and beyond. The long lead time required for the commercialization and widespread application of new technologies demonstrates clearly the need for a sustained research and development commitment, to ensure that advanced coal technologies are available to support coal's continued role in the energy systems of the next century.

Capital Investment in Coal

A substantial investment will be required to achieve the expansion in coal production, transport, and use projected in the

WOCOL study. Although data are limited, we have tried to estimate the order of magnitude of the aggregate capital investments required as well as the cumulative GNP and gross capital formation in the WOCOL countries in the OECD. This has allowed us to comment on the scale of capital investment in relation to capital formation in these countries.[10]

Ownership of mines, railways, ports, ships, and electric power stations varies greatly among countries. Many facilities are owned and operated by governments or other public authorities and are financed by governments directly or through sale of bonds of public utilities. Others are parts of the private sector and must satisfy their capital needs in domestic or international financial markets.

Capital Formation

For the 12 WOCOL countries in the OECD, we have estimated aggregate capital formation to year 2000, amounting to about $38,000 billion in U.S. 1978 dollars. This assumes that the 1973–1977 rates of capital formation reported by the International Monetary Fund (IMF) continue to the year 2000. These rates average out to about 23 percent of GNP for the 12 countries considered. It also assumes that GNP grows at the rates projected by the WOCOL country teams (see Appendix 1).

Capital Estimates for Coal Supply

We have found the concept of "coal chains" a useful device for visualizing the links in the system from mines to users and for illustrating the capital costs for each link as well as lead times for planning and construction of facilities. From the various examples in Chapter 8 (Figures 8-2 to 8-6) the data below has been assembled for the average capital costs in U.S. 1978 dollars per annual ton of new capacity.

- Mines $53   per annual tce
- Inland Transport $23   per annual tce
- Ports (Loading and Receiving) $23   per annual tce
- Ships $59   per annual tce

---

10. More detail on all these matters is found in Chapter 8.

Actual capital costs of coal supply chains will vary greatly among countries. The principal needs will be in coal producing countries and for ships. Investments by producers for mines, transport, and ports will be much greater than the investments by coal importers for ports and inland distribution systems.

Coal use in the WOCOL countries in the OECD is projected to increase by about 2,000 mtce from 1977 to 2000. Assuming that the average illustrative costs are broadly representative for other producers and exporters, this would require an investment for mines and internal transport of $76 per annual tce of new capacity, for a total of about $150 billion for the 1977–2000 period. The additional investment for export and import ports and ships to handle an increase in world coal trade of 600 mtce/yr would be $82 per annual tce or a total of about $50 billion. About three-quarters of this figure would be for the ships.

The total capital investment in coal supply chains in the OECD countries is therefore roughly $200 billion. Although such sums are large in absolute terms they are a very small proportion—less than 1 percent—of the aggregate capital formation of $38,000 billion estimated for these countries during the period to the year 2000.

Estimates for the countries outside the OECD are for an increase in coal use by the year 2000 of about an equal amount—2,000 mtce/yr, a large share of which is in the People's Republic of China and the Soviet Union. Estimates of total capital costs for mines, inland transport, and ports for these countries are not available but might be roughly comparable to the OECD costs, in order of magnitude.

Capital Estimates for Coal Use in Electric Power Systems

Most of the capital costs for implementation of a chain to mine, move, and use coal are accounted for by the large coal-using facilities such as power plants or synthetic fuel facilities. Typically, the capital costs for the power plant or conversion facility may be larger than those of the coal supply and transport infrastructure by a ratio of 3:1 or more. For example, if one assumes an average capital cost of $1 billion per 1 GWe of coal-fired capacity, this corresponds to a unit capital cost of about $500 per annual tce of coal use because 2 mtce/yr are required to fuel 1 GWe of electric capacity operating at a 65 percent capacity factor. This figure compares with our estimates of $76 per annual tce to supply the coal via mines and inland transport, and $82 per annual ton for ports and ships.

The WOCOL projections for Case B indicate that coal-fired power plant net additions for 1977–2000 in the OECD will amount to about 740 GWe. The aggregate capital cost would therefore be $740 billion using the $1 billion/GWe average cost assumption. This is a large sum, but financing it seems likely through a combination of domestic and international capital markets. Moreover, it is only 2 percent of the estimated aggregate capital formation during this period.

Factors Affecting Investment in Electric Power Stations

For many years electricity demand increased at an annual rate of 6–8 percent, and capacity was increased at a corresponding rate. Since the early 1970s, however, there has been a significant reduction in the rate of growth in some countries to half or less of the historic rate. Because many projects take up to 10 years to complete, installed capacity in many electric utility systems grew more rapidly than demand in the 1974–1979 period because of the completion of plants initiated in the earlier period of higher growth. As a result some electric systems had substantial excess capacity in 1979 that is likely to remain for some time as plants currently under construction including nuclear and coal plants are completed. Thus the need for new coal-fired power plants, and the associated need for capital to build them, is likely to be moderate through 1985 before accelerating thereafter, as we have reported in Chapter 2.

This situation, combined with a rather unfavorable financial environment in which some utilities are now operating, conflicts somewhat with the need for actions that can yield oil import savings in the early 1980s. In fact, electric utility boilers represent the largest potential market for early substitution of oil by direct use of coal. Utilities in the OECD currently burn 5 mbd oil, about two-thirds of which is consumed in the United States and Japan, and additional oil-fired plants are still under construction in some countries.

There exists significant potential for early reduction of oil use in these utility plants by "down-rating" or early retirement of older oil burning equipment, which may be an especially attractive option in countries with substantial excess electric capacity; by the conversion of existing plants to coal where possible; by the use of coal-oil mixtures (COM) if feasible where it is not economic to convert the existing oil equipment; or by replacing the existing oil-fired equipment with new coal-fired facilities. The doubling of world oil prices in 1979 makes all these coal substitution options considerably more economic than was previously the case. The possible oil savings from such coal substitution actions are summarized in Chapter 2 and considered in detail in the WOCOL country reports, which also provide age profiles on existing oil-fired electric capacity within each country.

However, even in cases where the costs of generating electricity from coal are known to be much less than from existing oil plants, many utilities are currently in a position whereby they have neither the need for new capacity nor the ability to raise additional capital. In such situations governments will need to provide financial incentives to encourage utilities to make the oil-saving investment decisions.

Potential Impediments to Financing Expanded Coal Capacity

There is substantial excess coal production capacity in some countries today. Short-term increases in demand can therefore be met in part by greater utilization of existing capacity, for example, production by the United States coal industry increased to 770 million short tons in 1979 compared to 654 million tons in 1978.

However, a major expansion of capacity will be required to meet the WOCOL projections of coal use in 1985–2000. Such an expansion will require substantial capital investments that will be made only if there is an assured and growing demand at prices that justify the investment in mines and transport to deliver coal to users. Many enterprises may not have the credit standing or operate in a sufficiently favorable political environment to encourage investors to undertake the long-range risks associated with loans so large that they can be repaid only over many years. For example, using an average figure for capital cost for mines of $53/tce/yr, a 1 mtce/yr mine requires an investment of $53 million, and the associated inland transport and ports for export may add another $33 million, making a total of an $86 million investment in the producing and exporting country. If it is a large 10 mtce/yr mine, the required capital is $860 million. Financing of actual projects of this size will involve many factors that must be resolved before financing is assured. Examples of the uncertainties involved are problems of delays, of escalation in costs, and of interruptions of various kinds that generally complicate the picture.

Another possible impediment revealed by our analysis is the relatively small size of the international capital markets and the heavy dependence on these markets by LDCs to fund their balance of payments deficits among other things. For this reason domestic capital markets should be used as fully as possible to finance coal expansion in the developed countries. Otherwise, LDC borrowers could be crowded out of the international markets they depend on by the large financing needs of energy projects in the developed countries.

Overall our analysis of capital markets and needs indicates that:

- The aggregate amount of capital required to finance the coal expansion projected is well within the capacities of capital markets, representing only a small fraction (about 3 percent) of the aggregate capital formation in the WOCOL countries. Individual projects may in some cases have some difficulties securing capital.

- Most of the capital for coal expansion is needed to build coal-using facilities.

- Lead times are long, and early decisions are required by consumers if the necessary coal supplies are to be available when needed.

- Provided that an adequate and stable framework is available, we are confident that the capital and resources needed by the coal market as envisaged in this study will be forthcoming. Governments can positively assist in the energy and coal market adjustment process by developing clearly defined energy and environmental objectives, and by adopting a consistent, efficient set of energy policies to achieve those objectives. This would include removing or modifying obstacles to coal production, international trade, and use that are not clearly in the long-term national interest.

Necessary Actions

The lead time for individual coal projects is long, usually 5–10 years. Even more significant, the large-scale industrial developments necessary to realize the coal expansion projected by WOCOL consist of many interweaving international coal chains that will take sustained effort over all of the next two decades to complete. They must be started now.

The problem is that decisions have to be taken well before the coal is actually needed, in the face of a range of uncertainties. These uncertainties cannot yet be resolved and, therefore, the prudent approach is to put the emphasis on insurance, with each country developing as far as economically practicable, the sources of energy it has available. In this way the security both of that country, and the world as a whole, is increased.

To this end, Europe can obtain greater security of energy supply by displacing oil imports both by building up its own production of coal where this is appropriate, and by increasing its imports of coal. These actions need not be in conflict; both are necessary. A complementary development of indigenous coal production and coal imports will lead to the maximum displacement of oil and increased long-term energy security.

In producing countries, major government decisions will be required to allow production and exports to reach the levels we have projected in this study. For example, in the United States many approvals by local, state, and federal agencies are required for the siting of new coal ports. In addition, there will be a need for an acceleration in the leasing of western coal land, as well as the construction of slurry pipelines to supplement railway expansion and carry part of the large volume of western coal to its markets.

Nations have different ways of achieving similar objectives because of the different roles played by governments in the coal and other energy industries. Thus the balance between legislative action and economic incentive required to facilitate the necessary expansion in the production and use of coal will vary between countries. International agencies such as the IEA and the World Bank can also assist. Many actors are involved and each has a part to play.

We believe that a broadly based national and international program of action is required to:

• Ensure that full and clear information on the urgency of the world's energy problem and on the essential role that coal must play is made available to governments, decision-makers, planners, and the public everywhere.

• Strengthen international cooperation in the development of expensive new technologies.

• Expand research and development into methods of environmental control of all aspects of coal production and use, and ensure that the results are applied to improve public acceptance of the expanded use of coal.

In particular, we believe that governments and public authorities have a major part to play and should give serious consideration to the following actions:

- define clearly their policies with regard to production and consumption of coal
- facilitate and/or establish, as far as they are involved, the necessary infrastructure for the increased production, transport, and use of coal
- avoid unnecessary delays in planning and licensing procedures
- facilitate the expansion of free international trade in coal
- promote, whenever they think it appropriate, a rapid substitution of oil by coal, both domestic and imported, especially in regions that are, or will be, heavily dependent on oil imports.

Coal can provide the principal part of the additional energy needs of the next two decades. But the public and private enterprises concerned must act cooperatively and promptly, if this is to be achieved.

BUILDING THE BRIDGE

Energy Units

Throughout this report we have used the unit *mtce* or *million metric tons of coal equivalent* as our standard measure of coal and energy use. This measure is based on the conventional unit of a *ton of coal equivalent* (tce), which is defined as a metric ton (2,205 pounds)* of coal with a specific heating value (7,000 kcal/kg or 12,600 Btu/lb).

Coals vary significantly in heat content, and most coals have a heat content of less than 12,600 Btu/lb. For this reason more than 1 metric ton of coal is often required to produce the energy content of 1 tce. In this regard it is important to recognize that it is physical tons that must be mined, transported, and burned or processed. The table below indicates the conversion factors from 1 tce to physical tons of coal of various calorific contents. In terms of oil equivalences, 1 tce converts to 0.65 tons of oil equivalent (i.e., 1 toe = 1.55 tce) and to 4.8 barrels of oil.

Conversion Factors for Coals of Various Calorific Contents

Type of Coal	Typical Calorific Content	Quantity Equivalent to 1 tce
Bituminous	12,000 Btu/lb	1.05 tons
Subbituminous	9,000 Btu/lb	1.4 tons
Lignite	7,000 Btu/lb	1.8 tons

---

* A metric ton is 10 percent heavier than a short ton (2,000 pounds), the unit commonly used in the United States.

One mtce provides 27.78 trillion (10^{12}) Btu's of energy. The following table provides an illustration of the amounts of energy provided by various quantities of coal.

Illustrative Scaling Comparisons for Various Quantities of Coal

Quantity of Coal	Indicator of Amount of Energy Provided
2 mtce	Annual primary fuel requirement for a 1,000 MWe electric power plant if it operates at a 65 percent capacity factor and generates 5.7 billion kWh per year electricity
5-7 mtce	Annual coal feedstock requirement for a 50,000 barrels per day synthetic liquids plant or a 250 million cubic feet per day synthetic gas facility
34 mtce	Amount of energy provided by 1 exajoule (10^{18} joules)
76 mtce	Amount of energy supplied annually by 1 million barrels per day of oil
100-140 mtce	Annual coal feedstock requirement for production of 1 mbdoe synthetic liquids

Comparison of the costs of various types of fuel is complicated by differences among the costs of fuel supply/delivery/use systems, by variations in fuel use efficiencies, and by the quality characteristics of different fuel types. The table below shows the costs of several fuels that are equivalent, on a calorific (Btu) basis only without accounting for the above differences, with oil at three price levels—$20/barrel, $30/barrel, and $40/barrel. For example, the table shows that coal at $142/tce is equivalent on a Btu basis to oil at $30/barrel.

Cost of Coal, Natural Gas, and Heat That Are Equivalent to Various Oil Prices

Fuel Type	Oil Price		
	$20/barrel	$30/barrel	$40/barrel
Coal ($ per tce)	$95	$142	$190
Natural Gas ($ per thousand cf)	$3.30	$5.00	$6.60
Heat ($ per million Btu)	$3.40	$5.10	$6.80

CHAPTER 1

WORLD ENERGY PROSPECTS

World Oil Supplies — Oil Supply Prospects in the Next Few Years — Oil Supply Prospects in the Medium and Long Terms — Oil Availability for Import by the OECD Countries — World Energy Consumption — Natural Gas — Unconventional Oil and Gas Sources — Nuclear Energy — Conservation — Renewable Energy Sources — Coal

World Oil Supplies

At the end of 1979, oil supplied more than one-half the total energy needs of the industrial nations of the Organization of Economic Cooperation and Development (OECD), and an even higher share of the energy consumed in the developing countries. Even in the centrally planned economy (CPE) countries, with their greater reliance on coal, oil provided about 30 percent of total energy supplies.

Because of its versatility, convenience of use, and—up to recently—its relatively low price, oil has become the world's most favored fuel. The rapid growth in world energy consumption over the last few decades has been largely supplied by oil. In the OECD region, oil use increased about 50 percent faster than total energy use between 1960 and 1978, and accounted for two-thirds of all the increase in energy use (Figures 1-1, 1-2). In Western Europe and Japan, oil provided more than 80 percent of the total increase in energy use, and the bulk of that oil was imported.

The ease with which oil can be converted to different products by refining has enabled it to satisfy a great diversity of markets. Many

The primary source materials for this review of world energy prospects are (1) *Energy: Global Prospects 1985–2000*, WAES (1977); (2) *World Energy: Looking Ahead to 2020*, World Energy Conference Conservation Commission, WEC (1978); (3) *Steam Coal Prospects to 2000*, International Energy Agency (December 1978); (4) *The World Oil Market in the Years Ahead*, National Foreign Assessment Center, CIA (August 1979); and (5) *Oil Crisis . . . Again*, British Petroleum Company Limited (September 1979).

Figure 1-1 Total OECD Energy Use (1960–1978)

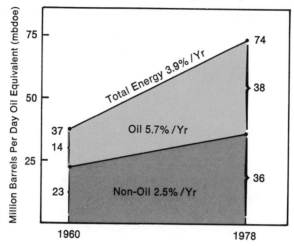

Figure 1-2 Oil's Share of the Increase in Energy Use— Total OECD and Selected Countries (1960–1978)

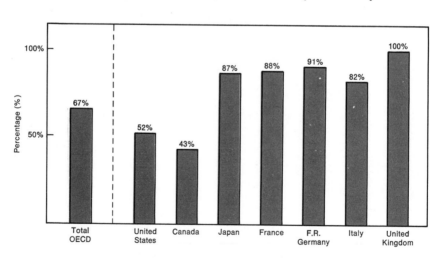

of today's energy markets owe their creation to oil's flexibility in this respect. Oil now provides all or most of the world's demand for transportation fuels and chemical feedstocks, as well as a significant share of the heat for homes, buildings, industry, and electric power plants. Because of its flexibility of use and the smoothness of its international supply and distribution system, it has also been used to meet surges in demand. Oil has played a vitally important role as a "swing" or "balancing" fuel in the energy systems of most countries.

It is with considerable concern, therefore, that we support the principal conclusion of nearly every recent world energy study: that *world production of oil from traditional sources is likely to peak before the end of the century and probably much sooner.* Of even greater consequence to most developed countries is that, after allowing for the increased consumption of oil by producer countries and the growing energy needs of developing countries, *the availability of oil for import to the OECD countries may already have peaked and will very likely be less in the year 2000 than today.*

These conclusions hold true in the different studies over the wide range of assumptions used about (1) the rate of world economic growth and world energy demand growth; (2) the oil production strategies of the Organization of Petroleum Exporting Countries (OPEC) member countries; (3) the future price of oil; (4) the rate at which new oil discoveries or improved production techniques will add to known oil reserves; (5) the contribution of alternative sources of energy; (6) the rate of oil production in non-OPEC countries such as Mexico, the People's Republic of China, and the Soviet Union; and (7) the growth of oil production from unconventional oil sources such as oil shale, tar sands, and heavy oil.

The leveling off of world production was recognized by the WAES Report (Figure 1-3), published in 1977, which stated: "The time when the production of oil will plateau and then decline is clearly in sight . . . if Saudi Arabia and certain other countries of the Arabian peninsula were to restrict production to only slightly above present levels, such a policy decision could lead to a leveling of world oil production in the early 1980s and thus to a failure to meet the projected demand much earlier than has been generally expected."

What in that report was considered to be a very pessimistic OPEC oil production assumption has turned out to be reality. World oil demand is now bumping against an OPEC oil production ceiling of approximately 30 mbd.

Figure 1-3 WAES Projection of World Oil Production (1975–2000)

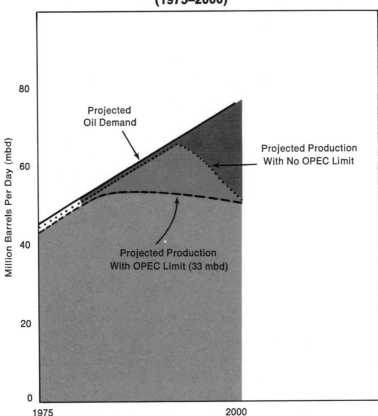

Note: Excludes oil production in CPE countries.
Source: *Energy: Global Prospects 1985–2000*, p. 129, Fig. 3-6.

Oil: A Historical Review

The two decades prior to 1970 represent the golden era of the oil-based world economy. Encouraged by low prices and expanding supply from many sources, world oil use increased rapidly. The OPEC member countries supplied nearly all of the increases in world oil supply. The price of oil, in real terms, was falling.

The situation changed in the early 1970s as oil demand began to approach the limits of supply capacity, with a resulting upward pressure on oil prices. One of the major factors was the rapid growth, during 1970–1973, of oil imports into the United States, which became the world's largest importer of oil (Figure 1-4). In 1971 the OPEC governments announced planned price increases amounting to

50 percent over a 5-year period. However, the Arab oil embargo initiated in October 1973 overtook this plan, and by January 1974 OPEC had virtually quadrupled its oil export price. (Figure 1-5).

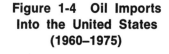

Figure 1-4 Oil Imports Into the United States (1960–1975)

Figure 1-5 World Crude Oil Prices (1960–1975)

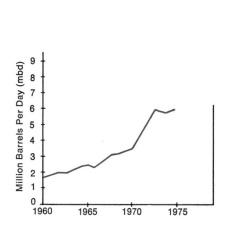

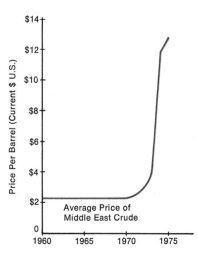

Note: Curve is smoothed.
Source: British Petroleum Co., Ltd.,
Oil Crisis . . . Again (1979), p. 5, Fig. 4.

Crude oil prices, in real dollar terms, subsequently declined slightly and remained below the 1974 level through 1978. Perhaps the most significant effect of the reduced pressure on world oil supply in this period was the complacency and skepticism induced in public attitudes toward the seriousness of the long-term world oil outlook. Modifications to oil consumption behavior patterns in most cases were rather ineffective. Valuable time was lost that might have been used to prepare economies for an adjustment to a limitation on world oil availability.

Events in the world oil market during and since the revolution in Iran in late 1978 have once again underscored the delicate balance between demand and supply for oil, as well as the sensitivity of the world economy to oil supply and oil price. Oil exports from Iran, previously about 5 mbd, were suspended from December 27, 1978 to March 4, 1979 before being resumed at a reduced rate of about 2–3 mbd. This reduction was partially offset by increased oil production and exports from Saudi Arabia and Kuwait. However, as in 1973–1974, the loss of several million barrels a day of oil from the world market for a few months caused oil importers to compete for the available oil. In an environment exacerbated by greater stock-building by consumers, spot market prices for oil climbed in early 1979 to $35–45 per barrel.

OPEC responded by rapidly adjusting its own prices upward. The 60 percent increase in oil prices announced by OPEC between January and June 1979 raised the official OPEC price to a range of $18–23.50/bbl depending on oil type. This increase more than offset the decline in real oil prices that had occurred in 1975–1978. Further price increases, to a range of $24–30/bbl, were announced by individual OPEC governments in mid-December 1979, and, for the first time in its history, OPEC was unable to agree on an official oil price structure. Still further price increases, to $26–35/bbl, were announced one month later. January 1980 oil prices were more than double the level of those of one year earlier, and 15 times higher than the level one decade earlier (Figure 1-6).

Figure 1-6   World Crude Oil Prices (1970–1980)

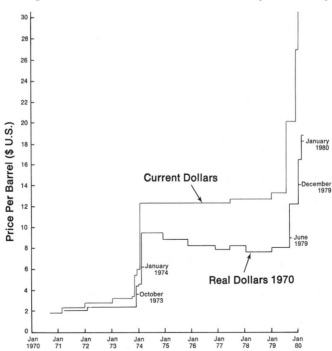

Oil Supply Prospects in the Next Few Years[1]

World oil production in 1978 (Figure 1-7) amounted to about 63 mbd; 30 mbd or about one-half was produced by OPEC. The balance was produced approximately as follows: OECD countries, 14 mbd; CPE countries, 14 mbd; and non-OPEC LDC countries, 5 mbd.

Figure 1-7 World Oil Production by Region in 1978

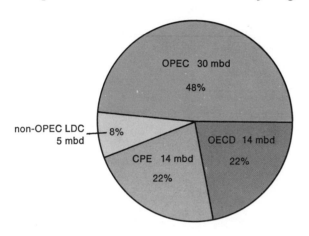

OPEC Countries

As early as 1976, seven OPEC governments representing more than one-half of all OPEC capacity announced oil-production target limits that, in total, were 5 mbd below their technical capacities. Table 1-1 shows that these OPEC countries were indeed implementing their stated production policies even before the Iranian revolution of 1978.

The declared policy of the members of OPEC is to limit their production individually to levels that in total amount to about 30 mbd. This approximates the present production level and compares with their estimated maximum possible capacity of 35 mbd (Figure 1-8). In fact, OPEC production has been roughly constant at 30-31 mbd since 1973.

---

1. Primary data for some of this discussion comes from *The World Oil Market in the Years Ahead* (1979). The detailed outlook for oil production and export-import balances, by country and region, is provided for reference in Table 1-A at the end of this chapter.

Table 1-1 Oil Capacity, Announced Production Limits, and 1978 Production of Several OPEC Member Governments (mbd)

Country	Estimated Usable Capacity at End 1975	Government Announced Production Limits as of early 1976	Actual 1978 Production
Venezuela	2.5	2.2	2.2
Ecuador	0.22	0.2	0.2
Libya	2.5	2.0	2.0
Qatar	0.7	0.5	0.5
United Arab Emirates	2.3	1.8	1.8
Kuwait	3.0	2.0	1.9
Saudi Arabia	10.8	8.5	8.3
Total	22.0	17.2	16.9

Source: WAES, except for 1978 actual production, which is from British Petroleum Co., Ltd., *Statistical Review of the World Oil Industry 1978.*

Figure 1-8 OPEC Oil Production and Capacity Trends (1970–1982)

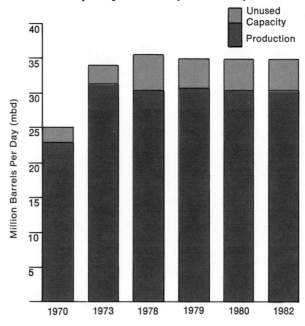

Source: *The World Oil Market in the Years Ahead* (1979).

Growing evidence suggests, moreover, that increased supply is not likely to be available. It is the policy of key producers such as Saudi Arabia and Kuwait to conserve oil in the ground as a better security for future generations than income they cannot spend or usefully invest. Reinforcing this conservation policy is the fact that by the upward adjustment of oil prices they can meet export revenue targets without the necessity of increasing or even maintaining export production levels (see Figure 1-9). What this means is that increased oil prices could lead to a fall in oil supply. For example, if Saudi Arabia has a hypothetical target export revenue of $65 billion per year, it can achieve this goal by exporting 9.5 mbd at $18.50/bbl or only 7.0 mbd at a higher price of $26/bbl.

Figure 1-9   The Effect of Oil Prices and Export Volumes on Possible Saudi Arabia Oil Export Earnings

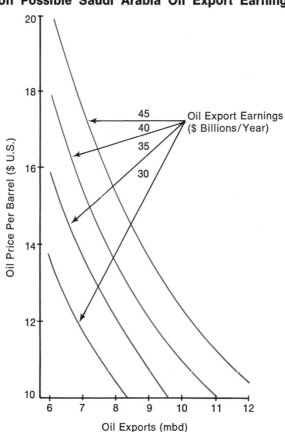

Note: Figure 1-9 shows combinations of oil prices and export volumes which could lead to constant export revenues of 30, 35, 40, and 45 billion dollars per year. Actual prices for Saudi Arabian oil increased to $26/barrel by January 1980.

Source: British Petroleum Co., Ltd., *Oil Crisis . . . Again* (1979), p. 17, Fig. 19.

These considerations, plus the fact that several other OPEC countries—for example, Algeria, Ecuador, Qatar, and Indonesia—have probably reached the plateau of their oil production because of resource depletion, have led most experts to conclude that the OPEC countries, as a group, are unlikely to increase their oil exports in the next few years. This is true even if Iran continues to export 2–3 mbd and there is no interruption in supply from other oil-producing countries in the Middle East.

If OPEC oil production is maintained at 30 mbd, exports from OPEC will decrease gradually because of the expected growth in their domestic oil use.

OECD Countries

Oil production in the industrial nations of the OECD is expected to increase slightly from 14.2 mbd in 1978 to about 15 mbd in the next few years and then decline. The primary reason for this increase is the rising rate of North Sea production in the United Kingdom and Norway. However, production in the United Kingdom is expected to reach a plateau in 1983 or 1984, and Norway continues to restrict its output. There is also a prospect of a decline in total North American oil production, as oil output from Prudhoe Bay in Alaska reaches its plateau.

Centrally Planned Economy (CPE) Countries

Oil production in the CPE countries, principally the Soviet Union and the People's Republic of China, amounted to 14 mbd in 1978. After meeting domestic demand and exports to other centrally planned economies, a balance of 1 mbd, mostly from the Soviet Union, has been exported to the world oil market.

The CPE countries are, however, expected by many analysts to cease being net exporters of oil by the early 1980s. In the late 1970s it appeared that oil production in the Soviet Union was beginning to level off, and there were indications that both oil production and oil export may decline in the early 1980s. The longer-term prospects for Soviet oil production may be more favorable, but they are not likely to be sufficient to meet the needs of the Eastern European area. Oil production in the People's Republic of China is expected to grow significantly, but most of the increase will be used to meet rapidly rising domestic needs.

70

Non-OPEC Developing Countries

The non-OPEC developing countries produced 4.6 mbd of oil in 1978 and consumed 7.4 mbd. Production in these countries is projected to increase to about 7.0 mbd by the early 1980s, with the major increase coming from Mexican production, which is projected to rise to 2.5 mbd or perhaps somewhat more, along with lesser increases in Egypt, India, Malaysia, and elsewhere. Much of the oil production increases in this group of countries will be consumed in the country where it is produced.

The growth in oil demand in developing countries that import their oil is being depressed by rising world oil prices. However, the rapidly expanding economies of countries like South Korea, Taiwan, and Brazil, as well as population increases elsewhere, are expected to cause some increases in total oil imports. Overall, roughly equal increases in imports and exports of oil by developing countries are expected.

Summary of Outlook for the Early 1980s

Total world oil production is thus expected to increase only very slightly in the next few years. However, the amount of oil available for import by the OECD nations may well decline over the same period because of increasing oil consumption within the oil-producing countries and less favorable oil balances in centrally planned economies. One outlook for oil up to 1982 is summarized in Table 1-2; this indicates that the availability of oil for import by the OECD could decline by 10 percent from 1978 to 1982.

Table 1-2 An Outlook on Oil Available for Import by the OECD Countries in 1980-82 (mbd)

	1978	1979	1980	1982
OPEC production	30.4	30.6	30.2	30.2
OPEC consumption	2.2	2.3	2.5	2.9
OPEC exports	28.2	28.3	27.7	27.3
Net imports of non-OPEC developing countries	2.8	2.8	2.7	2.6
Imports of non-OECD developed countries	0.4	0.5	0.5	0.6
Net exports of CPE countries	(1.0)	(0.8)	(0.4)	(—0.7)
Oil available for import by OECD countries	26.0	26.0	24.8	23.4

Source: *The World Oil Market in the Years Ahead* (1979).

Oil Supply Prospects in the Medium and Long Terms

There is a strong and growing body of opinion that OPEC production will remain at the current 30–31 mbd rate for a considerable time into the future, and few experts now expect OPEC production levels of much above 35 mbd for the remainder of this century. This outlook is in contrast to the widespread view, held as recently as 1977, that OPEC could produce 40–45 mbd during the 1980s and sustain that rate through the year 2000.

In addition to policy limits on oil production, the physical production potential in some countries has turned out to be less than expected. The most notable example is Saudi Arabia, where earlier estimates of a production potential of 20 mbd have been reduced to a maximum of 12 mbd, because of a reduction in the estimated oil recovery factor in major fields.

The supply of oil in response to any significant growth in oil demand is likely to remain at the discretion of a few oil-producing countries in OPEC, especially Saudi Arabia and Kuwait, which have large oil reserves capable of supporting expanded production. Figure 1-10 shows a recent update of world oil production prospects to the year 2000, and indicates by how much oil demand must be reduced in order not to bring pressure on this "discretionary" OPEC production. It requires a leveling and then decline in oil consumption. An increase in oil consumption of 3 percent per year from 1980 (less than half the pre-1973 trend) could be met only by a full expansion of OPEC discretionary production, and even then only until the mid-1980s. This plateau in world oil production is likely to exert constant upward pressure on oil prices over the medium and long terms unless action is taken to reduce world oil demand by a significant amount.

Oil Availability for Import by the OECD Countries

The WOCOL projected range for the future amount of oil available for import by the OECD nations is presented in Figure 1-11. It shows that available oil imports could be 15 percent less in the year 2000 than in 1979 if OPEC countries maintain production of 30 mbd (line II in Figure 1-11). This is true even with an optimistic assessment of oil supply and demand balances in non-OECD regions. If OPEC countries increase production to 35 mbd, then it may be

Figure 1-10 Illustrative Projection of World Oil Production (1980–2000)

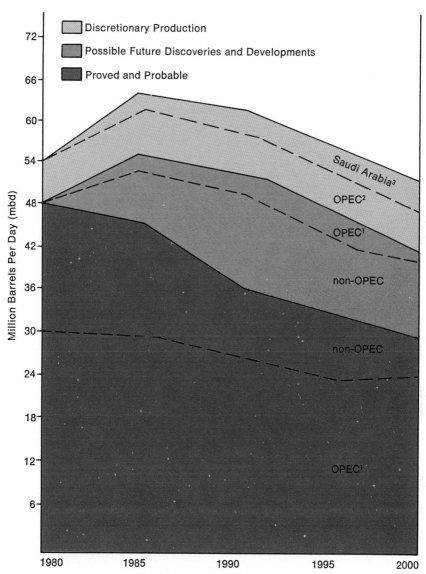

Note: Excludes oil production in CPE countries.
[1] Saudi Arabia 8.5 mbd, Iran 4 mbd, others at current production limits.
[2] Saudi Arabia 12 mbd, others + 2.8 mbd.
[3] Saudi Arabia 14–16 mbd.

Source: British Petroleum Co., Ltd., *Oil Crisis . . . Again* (1979), p. 19, Fig. 20.

Figure 1-11 Range of Net Oil Imports Available to OECD Countries (1960–2000)*

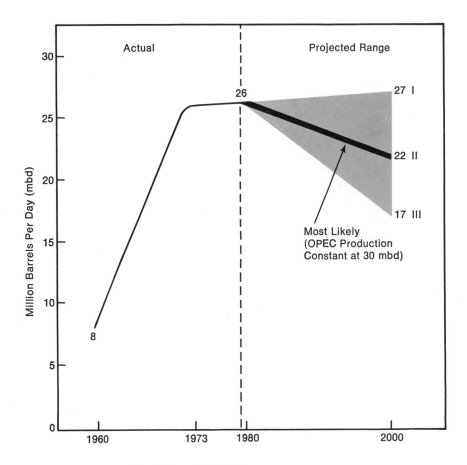

* Assumptions for Figure 1-11 (mbd).

	1978	Year 2000		
		I	II	III
OPEC Production	30.4	35	30	35
OPEC Consumption†	2.2	5	5	9
OPEC Exports	28.2	30	25	26
Net Imports of Non-OPEC developing countries	3.2	3	3	7
Net Import of CPE countries	−1.0	0	0	2
Net Oil Imports Available to OECD	26.0	27	22	17

† Includes bunkers sold by OPEC.

possible for the current level of OECD oil imports to be maintained (line I in Figure 1-11). However, less favorable assessments of oil balances in the non-OECD regions could result in oil imports to the

74

OECD falling below the 1979 level by one-third, even if OPEC produces 35 mbd (line III in the figure) and sustains it to the year 2000.

It is difficult to envisage any set of conditions that would provide for significant increases in OECD oil imports during this century. The more likely prospect, as shown, is for a gradual decline. This prospect contrasts with the pattern of 1960–1973, when OECD imports tripled from 8 to 25 mbd before leveling off.

Some of the implications of the oil limitation are reflected in a statement made by the International Energy Agency in Paris (May 1979): "Up until now, the expectation has too often been that at least into the medium term growing energy demand could be met by increasing oil supplies, principally from OPEC countries. But this can no longer be the case. In view of the limited possibilities for expansion of OPEC production, limited possibilities elsewhere, and the prospect of constant upward pressure on oil prices, it is now clearer than ever that continued dependence on oil to meet growing energy demand can no longer be viewed as a satisfactory option." [2]

In the face of limitations on world oil availability, it is clear that economic growth objectives can be reached only through expansion of energy sources other than oil and increases in the efficiency of energy use. Because the lead times necessary to expand energy supplies or to introduce effective energy conservation programs are generally long, the oil supply outlook leaves many countries with little prospect of being able to increase energy consumption in the short term. Over the longer term, alternative sources of energy will have to provide not only all the additional energy needed for future economic growth, but also the energy needed to offset any decline in oil availability. If these adjustments are not made, however, the penalty will be continuing stagnation in world economic activity.

World Energy Consumption

Total world energy consumption in 1978 amounted to about 125 mbdoe, about 63 mbd, or one-half, of which was supplied by oil. Supplies other than oil provided the other half—roughly divided into coal, 26 percent; natural gas, 17 percent; hydroelectric, 4.8 percent; and nuclear energy 2.4 percent, (Figure 1-12).

---

2. *Assessment of Medium and Long-Term Energy Outlook*, International Energy Agency (May 1979).

Figure 1-12   World Energy Supplies by Fuel in 1978

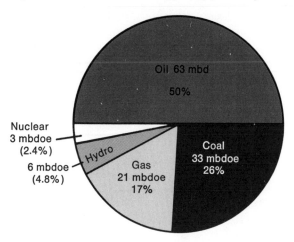

There is a need for substantial increases in world energy supplies over the next two decades even if the most ambitious efforts to improve the efficiency of energy use are successful. A 50 percent increase in world energy needs would be required by the year 2000 even if primary energy use grows only as fast as world population. Somewhat higher levels of world energy demand were projected for the year 2000 in the WAES (1977) and WEC (1978) studies.

Natural Gas

Natural gas currently provides about 44 trillion cubic feet per year of gas or 21 mbdoe of energy worldwide. It is a clean and convenient fuel, particularly favored for residential and commercial heating and for certain industrial applications. With populations and living standards continuing to rise, the demand for gas is likely to continue growing throughout this century.

World gas reserves are large and are capable of supporting increased world use over the next 20 years. However, many of the major reserves, for example, those in OPEC countries and in the Soviet Union, are distant from present and potential future markets. In some areas where natural gas is already used on a large scale the growth of production is being restricted by the slow rate of new discoveries, for example, in the United States and in parts of Western Europe. Many countries are relying on new sources of gas to provide stable or moderately expanded levels of supplies, including imports of liquefied natural gas (LNG), especially from OPEC countries;

production of low-, medium-, and high-calorific value gas from coal; and gas from unconventional sources such as geopressurized zones or tight formations.

Currently most natural gas is transported by overland pipeline. Underwater pipelines are also an option, for example under the Mediterranean from North Africa to Italy. To serve overseas markets gas can also be liquefied at very low temperatures and transported in special LNG tankers. Conversion to methanol is another option for overseas transport. The LNG and methanol systems both involve large capital investments in both gas-producing and gas-consuming countries. Shipping methanol does not require insulated tankers, but there is a greater energy loss in gas conversion to methanol with present technology (35 percent loss) compared with liquefied gas (25 percent loss).

The current volumes of liquefied natural gas in world trade are small, but are growing rapidly. In 1978, trade was about 500,000 bdoe; by 1985 this could rise to 1.5 mbdoe or more. LNG volumes of 5 mbdoe or more are technically possible by the end of the century. Such a level is equivalent to about one-quarter of total world gas production in 1979, and would require about 30–35 large LNG systems, each with a capital investment of around $4–5 billion.

Policies relating to natural gas vary widely among countries. Countries that are now large users of natural gas with large distribution systems and gas-using equipment in place, but with stagnating production, will have to determine the most attractive sources of supplemental gas or other energy sources to meet their future energy needs. Large gas producers, such as some OPEC countries, will have to consider the advantages of selling gas for export as LNG or methanol compared with conserving their gas reserves or using the gas domestically as a fuel or for fertilizers, petrochemicals, or iron and steel production.

The large world resources of natural gas would appear to indicate that supplies could be increased and that it would continue to make an important contribution to the energy needs of many countries. However, any substantial increase in world production and use of natural gas would depend on large exports from oil-producing countries in the Middle East and Northern Africa, exports that will not be forthcoming if these OPEC countries impose ceilings on natural gas exports similar to those now restricting oil exports.

Unconventional Oil and Gas Sources

Unconventional sources of gas in coal beds, shales, tight sands, and geopressurized formations may contain very large amounts of gas. A large resource base exists for unconventional oil in oil shale, tar sands, and heavy oil. Canada, for example, now produces about 7 percent of its oil from tar sands and plans a large expansion; however, difficulties in extraction and upgrading still hinder progress. The development of these sources is characterized by high capital costs and long lead times.

Interest in such unconventional sources has intensified recently as the outlook for conventional petroleum supply has worsened and energy prices have risen. There is a growing consensus that with increases in price, the world production of unconventional oil and gas could be increased from negligible levels in 1985 to about 3 mbdoe or perhaps somewhat more by the year 2000, and a considerable expansion after the year 2000 is projected by many experts.

Nuclear Energy

World nuclear power capacity today is about 110 GWe, and the completion of plants currently under construction would increase this figure to about 350 GWe by the late 1980s. Following the 1973–1974 Arab oil embargo, production of electricity from nuclear plants became a key feature of many national energy plans, and it is currently being relied upon by many governments to provide a large share of future energy needs.

Nuclear energy is technically capable of providing a large contribution to world energy supplies by the end of the century. The IAEA[3] has for example estimated that a world nuclear capacity of 1,030–1,650 GWe is feasible by the year 2000, that is, the equivalent of 27–43 mbdoe or 2,000–3,300 mtce/yr fuel requirements. The WEC in 1978 also projected that a capacity of 1,540 GWe could be possible worldwide by the year 2000.

However, such estimates of a 10–15 fold expansion of nuclear power in the next 20 years, though technically feasible, now seem unlikely. OECD projections of installed capacity in 1985 have been

3. "Nuclear Power Development: Present Role and Future Prospects," International Atomic Energy Agency, Vienna (August 1979).

reduced by 60 percent in the past 10 years (Figure 1-13). Projections of nuclear power expansion in some developing countries, for example Brazil, the Philippines, and Iran, have also been reduced.

Figure 1-13   Past Projections of OECD Nuclear Generating Capacity for Yearend 1985

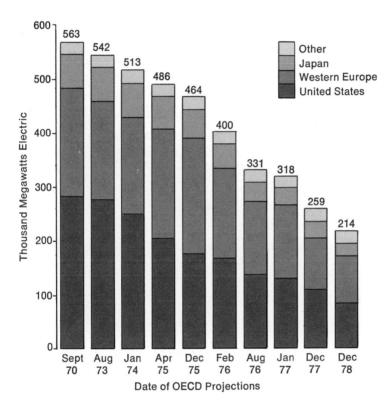

Source: *The World Oil Market in the Years Ahead*
(1979), p. 54, Fig. E-2.

Unless the uncertainties now affecting nuclear power in many countries are resolved soon, the world will need to plan for an energy future that can accommodate continuing delays in the expansion of nuclear power that would lead, among other things, to an increase in the requirements for coal. However, to the extent that there are advantages in diversifying electric fuel sources, coal and nuclear power can be considered to be complementary rather than competitive energy sources.

Conservation

The goal of energy conservation is to reduce the energy necessary to achieve any given level of economic activity. The future success of energy efficiency improvements will be central to achieving economic growth objectives as well as reducing the pressures on oil and other available energy supplies. Aggressive conservation programs now occupy a central place in the energy strategies of many nations. It has been widely suggested that by the year 2000 conservation could hold world energy consumption to levels 20–30 percent below what it would otherwise have reached. Even greater savings are suggested in some projections. Over the next 20 years therefore, conservation may well become one of the world's largest energy "sources."

Both in the short and the long runs, energy conservation is often the cleanest and cheapest way to respond to higher energy costs. Some actions can show rapid effects, for example, the installation of insulation in existing homes and office buildings. However, many years will be required for other measures to reach large-scale implementation. It takes at least a decade to change over a stock of motor vehicles, 20 to 30 years for most industrial equipment, and perhaps a century for a nation's entire stock of housing. The achievement of the largest savings will therefore tend to be gradual.

There is evidence that oil and other energy sources have been used more efficiently since the price increases beginning in the early 1970s. The response by industry has been most noticeable. For the OECD as a whole, the energy input per unit of industrial production has fallen by 7 percent below its 1970-1973 average. In the United States, industry's efficiency improvements have been even larger— about 20 percent (Figure 1-14). However, personal consumption of energy in vehicles, homes, and offices has not yet shown the same degree of efficiency improvement as in industrial energy use.

The projections made in the WOCOL country studies assume that high levels of energy conservation will in fact be achieved. The assumed savings lead to about a 25 percent decline in the amount of energy used in the OECD per unit of economic activity (GNP) by the year 2000. Vigorous efforts will be required to ensure that the conservation levels achieved are as large as possible.

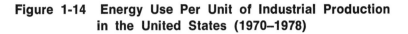

Figure 1-14   Energy Use Per Unit of Industrial Production in the United States (1970–1978)

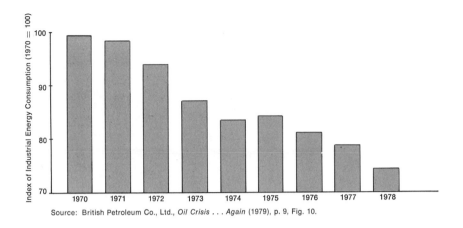

Source: British Petroleum Co., Ltd., *Oil Crisis . . . Again* (1979), p. 9, Fig. 10.

Renewable Energy Sources

Hydroelectricity is the only renewable energy source that currently provides significant amounts of commercial energy. It contributes about 20 percent of world electricity needs or about 5 percent of total world energy needs. It has many attractive features, and potential hydroelectric power developments have become increasingly competitive as oil prices have risen. The potential in the developing countries is particularly large, with the present development less than 30 percent of potential capacity as estimated by the World Energy Conference. Although the most favorable sites in the industrialized countries have already been developed, a large potential remains for better utilization of the existing sites, including low-head hydro and upgrading or rebuilding existing plants. The WEC projected that world hydroelectricity generation could more than double from 1972 to 2000 and would therefore continue to provide about 5 percent of the world's growing energy requirements at the end of the century.

Other renewable energy technologies—solar water and space heating, solar-generated electricity, photovoltaic energy conversion, biomass conversion, windmills, ocean-thermal energy, wave energy, tidal energy, and geothermal energy—provide little commercial energy today, but offer substantial promise for the future. Many of these technologies are at an early stage of development and significant breakthroughs are possible.

Geothermal sources provide useful quantities of energy in some areas today, and additional sources are being discovered and developed. The total contribution from geothermal sources will nevertheless inevitably remain geographically limited and relatively small in relation to world energy needs. Conversion of biomass to fuel alcohol is also under way in some countries, and is already making a useful contribution to transport fuel requirements in Brazil.

Solar energy has probably the greatest potential among the renewable energy sources. Solar collectors for domestic water heating and in some cases for space heating are already in use in warm climates such as that around the Mediterranean Sea. A penetration of the market to yield about 2 percent of total primary fuel usage in the year 2000 has been suggested, but this percentage could be somewhat higher. Solar electricity generation by photovoltaic or other means is not likely to become economic in time to make a large-scale contribution in this century.

Ultimately, renewable sources could become the principal source of world energy supplies. However, the relatively slow rate at which such new technologies can become economic and penetrate energy markets on a large scale is likely to limit the contribution of most of them to a small fraction of world energy needs over the next 20 years.

Three examples will serve to illustrate the limiting nature of penetration rates on the rapid commercialization of renewable energy technologies.

1. The turnover of an industrialized nation's housing stock is 50–100 years. Thus, progress in implementing solar heating in the industrialized world will be slow if applications of solar technologies are confined to new construction. To achieve a solar energy usage in domestic heating markets equivalent to 1 mbdoe (76 mtce/yr) in the United States would require the building of solar hot water and space heating systems in 2 out of every 3 homes built between 1980 and 2000. Such an effort would satisfy about 2 percent of the total U.S. energy needs in the year 2000.[4] However, the retrofit of existing homes, which is more costly than installations in new homes, could accelerate the introduction of solar home heating and increase its energy contribution.

---

4. *Energy: Global Prospects 1985–2000* (1977).

2. In the case of windmills, even with large-scale generators, electricity output is small. About 50,000 windmills with propeller diameters of 56 meters, and with a capacity of 1.5 MWe operating at 33 percent capacity factor, would be necessary to save the energy equivalent of 1 mbdoe.

3. In the case of biomass, it has been estimated that providing 1 mbdoe of alcohol from grain (as a supplement to gasoline use which was 7 mbd in 1979 in the U.S.A.) could require 90 million acres of land, which is about one-third of the cultivated land in the United States.[5]

There is, however, some reason to expect that the small-scale nature of many of the renewable technologies could allow them to penetrate markets somewhat faster than previous energy technologies, once they become economic. Hence, with vigorous support such as is now evident in some countries, some renewable technologies could be providing useful and rapidly growing amounts of energy by the turn of the century.

Coal

Coal now provides more than 2,500 mtce, or 33 mbdoe, of energy worldwide, more than any energy source except oil. Use is distributed 40 percent in the OECD region, 55 percent in centrally planned economy countries, and 5 percent in developing countries.

Coal will be required in the future to meet a major part of world energy needs, even after allowance has been made for a vigorous conservation program, development of solar and other renewable sources, continued nuclear growth, and the expected availability of oil and gas.

The size of the needs, the decisions that have to be taken, and the timing of such decisions are explored in the chapters that follow.

---

5. *Preliminary Forecast of Likely U.S. Energy Consumption/Production Balances for 1985 and 2000 by States,* U.S. Department of Commerce (1978).

Table 1-A An Outlook for World Oil Supplies
By Country and Region, 1978–1982 (mbd)

OPEC OIL PRODUCTION AND EXPORTS

	1978	1979	1980	1982
Crude Oil Production	29.7	29.8	29.3	29.1
Algeria	1.1	1.1	1.1	1.1
Ecuador	0.2	0.2	0.2	0.2
Gabon	0.2	0.2	0.2	0.2
Indonesia	1.6	1.6	1.5	1.4
Iran	5.2	3.0	4.0	4.0
Iraq*	2.5	2.9	2.4	2.4
Kuwait*	1.9	2.3	2.0	2.0
Libya	2.0	2.1	2.0	2.0
Neutral Zone	0.5	0.6	0.6	0.6
Nigeria	1.9	2.3	2.2	2.2
Qatar	0.5	0.5	0.5	0.4
Saudi Arabia*	8.1	8.8	8.5	8.5
UAE*	1.8	1.9	1.9	1.9
Venezuela	2.2	2.3	2.2	2.2
Natural gas liquids	0.7	0.8	0.9	1.1
Total OPEC Production	30.4	30.6	30.2	30.2
Less OPEC Domestic Consumption	2.2	2.3	2.5	2.9
Exports from OPEC	28.2	28.3	27.7	27.3

* Based on announced production policies.

CPE NET OIL BALANCE

	1978	1982
Net Exporters	3.2	2.0
USSR	3.0	1.7
People's Republic of China	0.2	0.3
Net Importers	—2.2	—2.7
Eastern Europe	—2.0	—2.4
Other	—0.2	—0.3
Export Balance	1.0	—0.7

OECD OIL PRODUCTION

	1978	1979	1980	1982
United States	10.3	10.2	9.9	9.2
Western Europe	1.9	2.4	2.9	3.9
Norway	0.4	0.4	0.5	0.7
United Kingdom	1.1	1.7	2.0	2.8
Other	0.3	0.4	0.4	0.4
Canada	1.6	1.8	1.7	1.7
Australia	0.5	0.5	0.5	0.5
Total OECD	14.2	15.0	15.0	15.3

NON-OPEC DEVELOPING COUNTRIES PRODUCTION AND CONSUMPTION

	1978	1979	1980	1982
Production	4.6	5.0	5.6	6.7
Consumption	7.4	7.8	8.3	9.3
Net Imports of Which:	2.8	2.8	2.7	2.6
Argentina	0.1	—	—	—
Brazil	0.8	0.9	1.0	1.0
Mexico	—0.3	—0.5	—0.7	—1.1
Peru	—	—	—0.1	—0.1
Egypt	—0.3	—0.3	—0.4	—0.7
India	0.4	0.3	0.3	0.4
Philippines	0.2	0.2	0.2	0.2
South Korea	0.5	0.5	0.5	0.7
Taiwan	0.3	0.4	0.4	0.6
Malaysia	—0.1	—0.1	—0.1	—0.1

Source: *The World Oil Market in the Years Ahead* (1979).

CHAPTER 2

ANALYSIS OF WORLD COAL PROSPECTS

Framework for the Analysis — Coal Use in the OECD Countries — Coal Use in Countries Outside the OECD — Total World Coal Use — World Coal Import Requirements — World Coal Export Potentials — International Coal Trade Patterns — National and Regional Implications of the Analysis

Framework for the Analysis

Country and Regional Study Approach

The geographic disaggregation scheme used by WOCOL to build the global analysis from the individual country studies and supplementary studies for seven other regions of the world is shown in Figure 2-1. Estimates of future coal use, production, import requirements, and export potential have been made for each of these regions.

Figure 2-1 WOCOL Geographical Regions

OECD Region
* Canada
* United States
* Denmark
* Finland
* France
* Federal Republic of Germany
* Italy
* Netherlands
* Sweden
* United Kingdom
* Japan
* Australia
  Other|OECD

Developing Regions
* India
* Indonesia
  East and Other Asia
  Latin America
  Republic of South Africa
  Other Africa

Centrally Planned
Economy Countries

* Poland
* People's Republic of China
  Soviet Union
  Other Centrally Planned

* Represented with members in WOCOL.

The analysis of coal demand by market sector focuses primarily on countries in the Organization for Economic Cooperation and Development (OECD), which represent 60 percent of the world's total primary energy consumption. The 12 OECD countries with Participants in WOCOL account for over 90 percent of total OECD energy use, gross national product (GNP), and coal consumption and production. The teams from the People's Republic of China, Poland, India, and Indonesia provided projections for their own countries. Estimates for the other centrally planned economies and developing countries were also made.

The assessment of world coal trade is based on coal import projections that were obtained from the country estimates of coal demand after subtracting domestic production. Additional information about world coal import needs was provided by special studies for the Western European nations not represented in the study, and for the developing countries, especially those of East and Southeast Asia.

Coal export potentials were estimated for each of the world's major coal-producing countries, most of which are represented in the study—Australia, Canada, Federal Republic of Germany, India, Poland, the People's Republic of China, the United Kingdom, and the United States. Current information for Republic of South Africa's export potential by the year 2000 was provided to WOCOL by that government's Department of Environmental Planning and Energy. Our estimates of coal export potentials of the Soviet Union, Colombia, and other possible exporters not participating in the study were based primarily on data published by the World Energy Conference and the World Bank.

Figure 2-2 illustrates the country study approach to estimating future coal use, production, and trade. The complete country and regional reports are contained in Volume 2, *Future Coal Prospects: Country and Regional Assessments.*

Coal Projection Methodology

The results of the analysis are reported in terms of four cases: A, A-1, A-2, and B. Each participating country team developed two reference projections describing its country's expected range of future coal requirements. The first projection, Case A, assumes a moderate rate of increase in the use of coal. The second estimate, Case B,

Figure 2-2 The Approach Used in the Country Studies

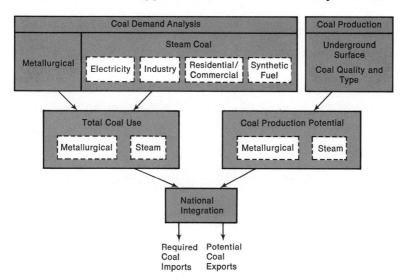

assumes a high rate of increase that would require levels of coal supply and demand close to their currently estimated upper limits of plausibility.

The estimates of total coal use for these two cases were developed through a disaggregated analysis of coal use in each country by market sector (metallurgical, electric, industrial, residential/commercial, and synthetic fuel). Estimates of the future electric utility market for coal were based on a range of assumptions about electricity growth rates and the expansion of nuclear power. Total electric generating capacity and fuel input balances were also developed. The range of estimates of coal use in the nonelectric markets, particularly in industry and as a feedstock for the production of synthetic fuels, was developed on the basis of existing projections, government policies to encourage the substitution of coal for oil and natural gas, and other factors. The growth of the steel industry and the changing steel-making technology were important variables in determining metallurgical coal requirements.

The estimates of coal demand were integrated into energy balances showing all fuels. Each country team formulated its own particular expectations of the general economic, technological, and political environment considered to be compatible with the estimates of moderate and high increases in coal use. Because of the long lead times required to build up the infrastructures for a large coal-using system, decisions taken in the next five years by coal users to build new coal-using facilities will largely determine the evolution of coal demand in the 1985–2000 period. Accordingly, we have based our coal demand projections on current expectations of coal use, which vary from country to country, rather than on postulated future economic states of the world. No attempt was made to develop a general global economic framework for the next 20 years for the study.

In general, the projections of growth in coal use would be decreased by lower rates of growth in the economy, in total energy demand, and electricity demand in particular, whereas the estimates of coal demand would be increased by lower estimates of oil and gas availability, nuclear power expansion, and development of alternative sources of energy. Higher prices for oil and gas would also encourage the accelerated substitution in favor of coal. Appendix 1 provides the summary coal and energy data used for WOCOL Cases A and B for countries participating in the World Coal Study.

Our review of the total OECD projections for oil imports and the expansion of nuclear power, made on the basis of the individual country estimates, led to development of the two additional variations on Case A, that is, A-1 and A-2. Case A-1, involving oil limitations, superimposes on Case A the incremental coal demand emerging from the country responses to an assumed 20 percent reduction in total oil use below the reference projections for the year 2000. This would reduce projected OECD demand for oil imports to a level consistent with our estimates of what is likely to be available. Individual responses to this situation were not necessarily limited only to bulk substitution of coal for oil but ranged over the whole spectrum of possible actions including increased conservation, increased use of nuclear power, increased supplies of natural gas and accelerated development of renewable energy sources. Nevertheless, the demand for coal obtained in Case A-1 is substantially higher than in Case A. Case A-2, introducing nuclear delays, superimposes on Case A-1 the additional requirements for coal that would develop if coal were substituted directly for a 30 percent reduction below the reference projection of total OECD[1] nuclear capacity in the year 2000.

---

1. Although a figure of 30 percent was used for the OECD as a whole, assessment of possible nuclear delays varied among WOCOL countries (see Volume 2).

Responses to limitations on oil supplies and delays in nuclear power expansion may be expected to be complex and to vary significantly from country to country. The country reports in Volume 2 illustrate this diversity. The effect of possible oil and nuclear energy supply limitations are reported here only for Case A. The original estimates of the high coal increase in Case B are generally near the upper limits now considered plausible. However, some teams have looked at the implications of limitations on oil supply and delays in nuclear power expansion also under the circumstances of high rates of growth of coal use (Case B), and these are reported in Volume 2.

Coal Use in the OECD Countries

Figure 2-3 shows the projected coal use and market distribution in the OECD[2] countries to the year 2000. The increase in total coal use in both cases is substantial, by about 1,000 mtce/yr or a doubling in Case A, and by 2,000 mtce/yr or a tripling in Case B.

Figure 2-3 Coal Use and Markets in the OECD (1977–2000)

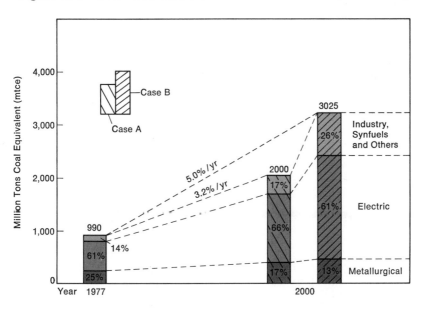

---

2. Throughout this chapter OECD totals would be increased slightly with the inclusion of New Zealand. The IEA Steam Coal Study estimated that coal use in New Zealand would increase from 2.4 mtce in 1976 to 7.7 mtce in 2000, with all the coal produced domestically.

The major coal use today and in the year 2000 in both cases is in electricity generation, which represents more than 60 percent of the total coal use. The high growth projected for coal use in non-electric energy markets, especially industrial and synthetic fuel uses, accounts for the growth of these markets from 14 percent of total OECD coal use in 1977 to a range of 17 to 26 percent in 2000. Metallurgical use of coal grows more slowly, and its share in the coal market drops from 25 percent in 1977 to 13–17 percent by the end of the century.

The Electric Power Market

The use of coal in coal-fired power plants is the only market for coal that has seen significant growth over the past two decades. The dominance of electric utilities as the driving force for increased coal use will continue for the next several years and probably for most of the 1990s as well. Figure 2-4 shows the required coal-fired capacity for the total OECD.[3] The figure indicates that even in Case A, where electricity demand grows moderately, coal-fired capacity in the OECD more than doubles from 350 GWe to 825 GWe by the year 2000. Coal capacity in Case B increases further to 1,090 GWe, as a result of an electricity growth rate about 1 percent higher than in Case A.

Figure 2-4 OECD Electric Capacity Mix (1977–2000)

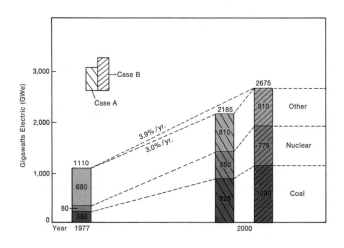

3. Each GWe of electric capacity requires 2 mtce/yr coal if the plant is operated at a 65% capacity factor.

Regional variations in the projections for the use of coal in the electric power market are indicated in Figure 2-5. Penetration of coal in the electric capacity mix is projected to be greatest in OECD regions with substantial coal resources. In North America and Australia there is a significant increase in coal's contribution to electricity capacity above 1977 levels, whereas in Europe virtually no change is projected—although, of course, the absolute level of coal-fired capacity does increase substantially. The most significant changes relative to today are projected to occur in countries with little or no exploitable coal resources and where coal presently provides only a small contribution to electric power generation. In Japan, for example, coal capacity increases from 4 percent of the total capacity to 15 percent by the end of the century. As another example, in the Netherlands coal is projected to increase its contribution from 5 percent today to 70 percent in 2000 in Case B. Increased coal requirements in the electric sector under conditions of a limitation on oil availability or delays in nuclear power expansion have the effect of increasing the coal penetration even further.

Figure 2-5 Coal-Fired Percentage of Electric Capacity— Total OECD and Selected Regions (1977–2000)

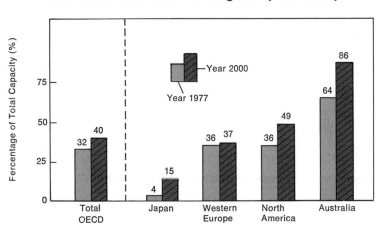

Figures for the year 2000 are for Case B (high coal increase); penetration for Case A is very similar.

Table 2-1 indicates the range of projections for coal use in the electric market for each of the WOCOL countries and for the OECD as a whole. Coal use in OECD electric power stations is projected to increase from 600 mtce in 1977 to a range of 1,325–1,850 mtce/yr by the end of the century. In addition to the expansion of U.S. coal use, it is noteworthy that Australia, with its small population, is projected to be the second largest user of coal for electricity generation of all OECD countries by the year 2000. The projected rate of growth of OECD coal use is moderate through 1985, but conversion of oil-fired power plants to coal, either directly or by use of coal-oil mixtures if feasible, or down-rating or early retirement of old oil plants could accelerate the pace of coal expansion in the 1980s.

Table 2-1 Coal Requirements in the OECD Electric Markets (mtce)

Country/Region	1977	1985		2000	
		Case A	Case B	Case A	Case B
Canada	17	31	28	50	65
United States	372	500	560	800	1170
Denmark	3.8	9.9	9.9	8.6	19
Finland	1.9	1.1	1.1	3.2	7.9
France	22	12	25	10	45
Germany, Fed. Rep.	61	76	79	99	106
Italy	1.8	10	10	17	37.5
Netherlands	1.1	3.9	3.9	13.7	25.9
Sweden	0	0.4	0.7	6.7	15.2
United Kingdom	65	68	72	70	77
Other Western Europe	24	26	38	85	98
Japan	6	13	13	57	72
Australia	25.0	42.9	42.9	107	107
Total OECD*	600	800	880	1325	1850

* Totals are rounded.

The Metallurgical Coal Market

Use of coal to produce coke for metallurgical purposes is the second largest OECD consumer of coal today, accounting for about 250 mtce in 1977 or 25 percent of total coal use. This figure has grown only moderately since 1960, when metallurgical use stood at 190 mtce. Our projections indicate a moderate growth to 330-375 mtce/yr by the end of this century, when coal for metallurgical use in the OECD will decline to about 15 percent of total coal use.

Despite the moderate growth projected in OECD coking coal needs, growing steel industry requirements outside the OECD may result in a tightening of the world markets for high-quality (low-

94

volatility) coking coals. This concern has led to intensive research efforts in many countries to commercialize formed coke production, which would allow use of noncoking or weakly coking coals for the manufacture of coke. About 10 different processes are now under development in pilot plants throughout the world. In addition, expansion of direct reduction processes to produce iron could further decrease future requirements for coking coals.

The Industrial Market

The use of coal in industry has declined sharply over the last two decades and accounted in 1977 for only about 90 mtce or 9 percent of total OECD coal use. Nearly all WOCOL teams expect this trend to be reversed. In our projections industrial coal use in the OECD expands by a factor of at least 2 and possibly by 4 (Table 2-2). Countries projecting the largest increase in this market for coal are Canada (from 1 mtce in 1977 to 11–20 mtce in 2000), the United States (from 60 to 125–220 mtce), France (from 3 to 14–40 mtce), and the United Kingdom (from 9 to 28–47 mtce).

Table 2-2 Coal Requirements in OECD Industry Markets (mtce)

Country/Region	1977	1985		2000	
		Case A	Case B	Case A	Case B
Canada	1	4	4	11	20
United States	60**	70	80	125	220
Denmark	0.7	0.7	1.0	0.7	1.4
Finland	0.7	0.9	0.9	1	1
France	3	5	12	14	40
Germany, Fed. Rep.	5	6	8	8	12
Italy	0.2	1	1.5	2.5	3
Netherlands	0.3	2.2	2.2	3	4
Sweden	0.3	1.8	1.8	3.3	3.3
United Kingdom	9	12	16	28	47
Other Western Europe	4	5	6	7	13
Japan	4	4	5	7	12
Australia	4.4	11.1	11.1	16.1	16.1
Total OECD*	90	125	150	225	400

* Totals are rounded.
** Includes 7 mtce residential/commercial use.

As in the electric market projections, the major expansion in industry markets is expected to be realized after 1985. The lead times required to convert existing equipment or install new coal-fired equipment in industry mean that the 1985 energy use pattern is largely determined by actions that have already been initiated. This market is however projected to grow quite rapidly at 5–7 percent per year during the 1985–2000 period.

Those industries in which coal use is expected to increase significantly include cement, chemicals, petroleum refining, and pulp and paper manufacture. The conversion of coal to low- or medium-Btu gas may have a special attraction to certain sectors of the industrial market that need fuel in gaseous form. In some countries the commercialization of fluidized bed combustion, eliminating the need for installation of flue gas desulfurization devices, would also tend to increase the use of coal by industry, as could the use of coal-oil mixtures (COM) if this proves feasible.

However, the use of coal by industry, particularly by small firms, is dependent on resolving the current reluctance of individual manufacturers to turn away from using oil or natural gas and to face the uncertainties of a new coal distribution system, an unfamiliar coal use technology, and more complex environmental requirements.

The Synthetic Fuel Market

A significant new market for coal feedstocks for synthetic oil and gas plants is anticipated to develop in some countries in the 1990s, encouraged by rising world oil and gas prices and governmental incentives. The OECD market for coal synfuel feedstocks by the year 2000 is estimated to range from 75 to 335 mtce (0.6–2.8 mbdoe products). This would require the construction of about 15 large-scale synfuel plants (@ 50,000 bdoe) this century in Case A and 67 plants in Case B.

The Case A estimates of synthetic fuel development could be viewed as a demonstration program that would have little impact on OECD energy supply and demand balances during this century. The Case B synfuels programs would begin to have a significant effect in several countries in the mid-1990s and could reduce the need for OECD oil imports by nearly 3 mbd in 2000 (12 percent of current oil imports). In both these estimates, two-thirds of the total synfuel capacity in OECD countries is projected to be built in the United States. Other countries that project potentially significant markets for coal for synfuels by the end of the century are Japan, Australia, Canada, the Federal Republic of Germany, and France.

The Residential/Commercial Market

Coal has largely disappeared as a fuel for homes and commercial facilities in the OECD, with use in 1977 being less than 50 mtce. This compares with 5 times as much, or nearly 250 mtce, as recently as 1960. The reasons that led to this decline include the lower costs of oil and gas up to 1973 and the added convenience and cleanliness of handling such fuels.

Although the costs of oil and gas have been rising sharply since 1973, the convenience differential still exists, and nearly all country teams project that the use of coal in homes and office buildings will remain insignificant in this century. It has been suggested, however, that a revival of interest in coal as an energy alternative for homes and office buildings cannot be ruled out; and indeed there are some indications that this is already happening. For example, such coal use is projected to grow from 10 mtce in 1985 to 15–21 mtce by the year 2000 in the United Kingdom. The domestic heating market could be encouraged in some countries by the use of centralized district heating or cogeneration schemes. Alternatively, smokeless coal burned in domestic furnaces or stoves could become an attractive alternative to oil or gas furnaces, or wood stoves, in cold regions. Perhaps the most convenient way for coal to enter this market would be indirectly through liquefaction and gasification where and when these processes become technologically and economically feasible.

Total Coal Use in the OECD

The projected total coal use in the OECD, by country, is indicated in Table 2-3. Annual steam coal requirements grow from about 750 mtce in 1977 to a range of 1,670–2,650 mtce in 2000. The associated growth rates are in the range of 3.5–5.6 percent per year. Adding the requirement for metallurgical coal, total OECD coal use grows to 2–3 billion tce by the year 2000.

The growth of steam coal use is projected to be very rapid for many countries, in some cases reflecting the low current use. Canada, Italy, the Netherlands, Sweden, and Australia all project more than a 4-fold increase in their steam coal use in both the moderate and the high growth cases. Whereas in France coal use is expected to triple (under Case B), in Japan an increase of 6–13 fold

97

Table 2-3 Total Coal Requirements in OECD Countries (mtce)

Country/Region	STEAM COAL					TOTAL[a] COAL				
	1977	1985		2000		1977	1985		2000	
		Case A	Case B	Case A	Case B		Case A	Case B	Case A	Case B
Canada	18	35	32	67	106	25	44	41	82	121
United States	432	570	640	975	1,590	509	655	725	1,075	1,700
Denmark	4.6	10.7	11.1	9.4	20.9	4.6	10.7	11.1	9.4	20.9
Finland	3.5	3.4	3.4	8	12	4.3	4.4	4.4	9	13
France	31	19	42	31	105	45	35	59	48	125
Germany, Fed. Rep.	79	93	102	125	153	102	119	126	150	175
Italy	2.4	11.3	11.9	19.5	48.5	13.5	22.3	22.9	31.5	60.5
Netherlands	1.5	7	7	20	34	4.5	10.4	10.4	23	38
Sweden	0.3	2.9	3.2	14.3	23.1	2.1	5.1	5.4	17	26
United Kingdom	91	90	98	117	158	109	107	115	133	179
Other Western Europe	33	36	51	96	121	51	61	83	135	175
Japan	10	17	18	64	132	79	97	102	150	224
Australia	29.7	55	55	124	149	38	65	65	141	166
Total OECD[b]	740	950	1,075	1,670	2,650	990	1,235	1,370	2,000	3,025

[a] Total includes steam plus metallurgical coal.
[b] Totals are rounded.

is projected. In the United States coal use grows to between 2 and 3 times 1977 levels of consumption.

Coal's Share of Total OECD Energy Supplies

Figure 2-6 shows coal's growing contribution to total OECD energy supplies, based on the country team estimates of primary energy requirements and supplies of oil, gas, and nuclear as well as hydroelectric, solar, and other renewable resources. As indicated, coal's share of total OECD energy supplies has fallen from 37 percent in 1960 to 18 percent in 1977. Both cases project that this trend will be reversed. Coal's share rises by the end of the century to 25 percent in Case A and 32 percent in Case B.

The figure also indicates that the Case A estimates are associated with a very moderate growth rate in OECD primary energy

Figure 2-6   OECD Primary Energy Requirements by Fuel (1977–2000)*

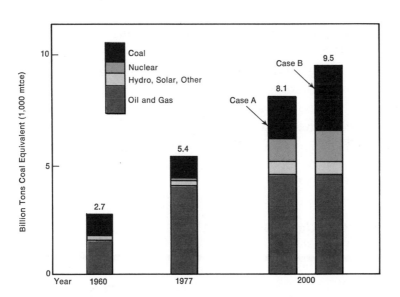

* Assumptions for Figure 2-6 (billion tce)

	1960	1977	2000 Case A	2000 Case B
Coal	1.0	1.0	2.0	3.0
Nuclear	0	0.15	1.0	1.4
Hydro, Solar, Other	0.2	0.25	0.6	0.6
Oil and Gas	1.5	4.0	4.5	4.5
Total Primary Energy Consumption	2.7	5.4	8.1	9.5

consumption of 1.75 percent annually for the 1977–2000 period. The Case B estimates imply an energy demand growth rate of 2.5 percent per year. Both these growth rates are lower than the 2.7 percent per year assumption of the IEA steam coal study (December 1978) and are significantly below the 2.1–3.0 percent per year range projected for OECD by the WAES study (1977). Our estimates reflect a growing awareness of the effects of rising energy prices and improved energy efficiency in slowing energy growth.

A significant regional variation is apparent in the country estimates of energy demand, as might be expected. Japan and Australia project growth rates higher than most of the other OECD countries, as indicated by the following range of average energy growth rates for the 1977–2000 period for Cases A and B:

North America	1.3–2.3%/yr
Western Europe	1.8–2.4%/yr
Japan/Australia	3.1–3.6%/yr

Our estimates of the growth in energy demand are compatible with OECD economic growth rates of 3–3.5 percent per year if the strong energy conservation assumptions in the individual country projections are realized. For example, the Case A estimates imply that OECD energy use grows little more than half as rapidly as economic activity, compared with an average one-to-one relationship in the 1960–1973 period. Another way of illustrating this is that the ratio of energy use to GNP (1978 dollars) falls in Case A from 1.00 tce/$1,000 in 1977 to 0.74 tce/$1,000 in the year 2000.

The Effect of Limitations on Oil Imports and Nuclear Delays on OECD Coal Needs

The individual country assumptions made in both Case A and Case B about future oil import needs add up to a total OECD oil import requirement of 29 mbd by the year 2000. This is an increase by about 12 percent or 3 mbd from the 1977 level of 26 mbd. Such an estimate includes projections that the United States will limit its imports to the current level of 8–9 mbd; that Japan's oil imports will increase from 5.3 mbd in 1977 to about 7 mbd; and that net imports to Western Europe will hold roughly constant at 12–13 mbd.

However, even this relatively small growth of 3 mbd in oil imports exceeds our most optimistic estimates of the oil that will be available for import. As shown in Chapter 1 (Figure 1-11), we expect that future oil supplies available for import to the OECD will be no higher than the levels in 1977, and they are likely to be less.

Figure 2-7 presents the possible impact of the projected oil limitation on the OECD coal requirements. Case A-1 is obtained by superimposing on Case A the individual country responses to a 20 percent reduction in total oil use below their original projections for the year 2000. This would reduce the projected OECD demand for oil imports in the year 2000 to about 21–22 mbd, which is consistent with what we consider likely to be available at that time. The responses show an increase in OECD coal use of about 500 mtce in the year 2000, along with other measures to reduce oil demand and increase alternative energy supplies.

Figure 2-7 The Effects of Oil Limitations and Nuclear Delays on OECD Coal Requirements (1960–2000)

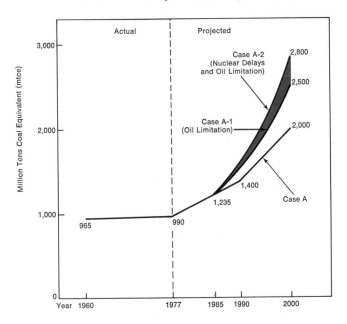

An example of possible responses to an oil import limitation is illustrated in the United States report in Volume 2. The Case A-1 import reduction is 3.4 mbd (from 8.4 to 5.0 mbd) in the year 2000. This is assumed to be achieved by four measures:

(1) 1.5 mbd — reducing oil use in utility boilers from a projected 1.8 mbd in 2000 to 0.3 mbd;

(2) 0.5 mbd — substituting coal for oil in industry boilers;

(3) 0.5 mbd — increasing supplies of unconventional gas by 1 tcf/yr above Case A levels; and

(4) 0.9 mbd — expanding the Case A coal-based synthetic fuels program from 0.4 to 1.3 mbdoe by year 2000.

Many countries have been depending on an expansion of nuclear power to provide a significant share of their incremental energy needs for the rest of this century, and this is evident in the reference WOCOL projections. For example, the Japanese projections assume an expansion of nuclear capacity from 8 GWe in 1977 to about 100 GWe in 2000; those for France from 4.6 to about 80 GWe; for the Federal Republic of Germany from 7.4 to about 60 GWe. These countries currently expect that nuclear energy will provide a larger share of their increased energy needs during this century than coal.

A 7–10-fold expansion of total OECD nuclear energy capacity is implied by the original WOCOL projections, from about 80 GWe in 1977 to 550–775 GWe in 2000. Realization of these projections, however, depends on the resolution of the uncertainties that are now delaying the development of nuclear power in many countries.

The second variation (Case A-2) shown in Figure 2-7 superimposes on Case A-1 the increases in coal demand arising from the assumption of delays in the expansion of nuclear power, over and above the coal increases caused by the oil limitation. A reduction of 30 percent in the Case A projections of total OECD nuclear power capacity would lead to a Case A-2 nuclear capacity of 400 GWe instead of 550 GWe in the year 2000 and, superimposed on Case A-1, an additional increase in OECD coal use of 300 mtce to a total of 2,800 mtce if coal were to be directly substituted for the reduction in nuclear capacity. The combined effect of the limitations on oil and of nuclear energy delays increases the Case A-2 estimate of OECD coal use up to nearly the original high coal level of Case B, which was 3,025 mtce in year 2000.

Coal's Share in the Increase in OECD Energy Supplies

Figure 2-8 shows the large part that coal will supply toward meeting increased OECD energy needs over the next 20 years. In Case A, the share for coal is 37 percent. In the assumed oil limitation in Case A-1, OECD oil use declines, whereas coal use grows to supply 55 percent of the increase in OECD energy needs. Possible delays in OECD nuclear expansion, in combination with the oil limitation, accounts for an additional rise to 67 percent in Case A-2, while the nuclear portion falls from 32 to 20 percent.

Figure 2-8   Coal's Share in Meeting the Increase in OECD Energy Needs (1978–2000)

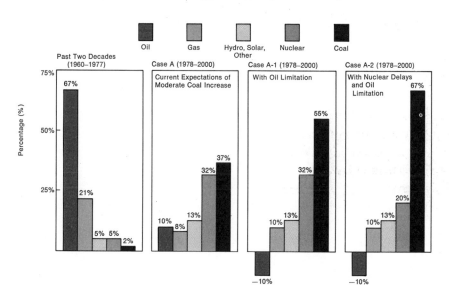

Coal will need to provide half or more of the total increase in energy needs over the next two decades, while oil provides little or none of the increase. This contrasts sharply with the past two decades, when coal provided virtually none of the increase, and oil provided two-thirds of the increase.

The significance of coal in providing a major share of the increased energy needs of many nations is shown in Figure 2-9.

The Acceleration of OECD Coal Demand

The WOCOL projections show that the rate of growth in OECD coal demand is likely to accelerate sharply in the mid-to-late 1980s. The projections show a moderate rate of increase in demand

Figure 2-9 Coal's Share in Meeting the Increase in Energy Needs—Total OECD and Selected Countries (1978–2000)

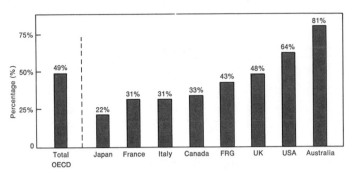

Figures are for Case B. Corresponding figures for Case A are OECD, 37%; Japan, 14%; France, 2%; Italy, 13%; Canada, 27%; Federal Republic of Germany, 25%; United Kingdom, 44%; the United States, 65%; Australia, 65%.

of 3–4 percent per year over the 1980–1985 period. They indicate that beyond 1985, as national actions to substitute coal for oil begin to take effect and as the need for new coal-fired power plants becomes greater, the rate of growth of demand for coal will increase rapidly.

If, however, this acceleration does not take place, and the moderate rate of coal expansion simply continues, coal's contribution in the OECD countries will be limited to about 2 billion tce by the end of the century—the Case A level. Such a contribution, even though large, would be inadequate to compensate for the projected leveling of world oil supplies and possible delays in nuclear expansion programs, even with the lowest levels of energy demand growth that we have investigated.

Because of the substantial lead times necessary for expanding coal production and use, the WOCOL projections of an accelerated coal expansion incorporate the assumption that many actions are initiated in the early 1980s to build new coal-using facilities, to convert existing oil-using facilities to coal-firing (especially cement kilns and power plants) or to develop the use of coal-oil mixtures, and in some cases to order the construction of coal-based synthetic oil and/ or gas plants. The assumed actions are described fully in the WOCOL country reports in Volume 2.

Coal Use in Countries Outside the OECD

Countries outside the OECD used 1,500 mtce coal in 1977, or 50 percent more than OECD countries. The largest consumers were the Soviet Union (490 mtce), the People's Republic of China (368 mtce), Poland (127 mtce), India (72 mtce), and South Africa (61 mtce). Although estimates of their future coal requirements and production are less detailed than those for OECD, it is apparent that these countries will experience many of the same pressures on energy supplies as countries in the OECD.

The WOCOL country team estimates for India and Indonesia, taken together with the special regional analyses carried out by WOCOL, indicate a large increase in coal use in the developing countries—from 150 mtce today to a range of 600–900 mtce by the year 2000. India alone projects an increase from 72 mtce in 1977 to 280 mtce by 2000. Indonesia projects a growth from less than 1 mtce in 1977 to about 20 mtce in 2000.

The largest coal use will be for electric power generation in those developing countries with large rates of economic growth that do not have abundant unexploited hydroelectric resources (e.g., South Korea, Taiwan, the Philippines, and some countries of Latin America). Our estimates indicate that steam coal requirements for the developing countries of East and Southeast Asia[4] will be at least 70 mtce in the year 2000, and could be as high as 190 mtce, compared to less than 20 mtce today. In addition, there is an estimated 3–4-fold increase in coal use in the People's Republic of China to more than 1,000 mtce by the end of the century.

Coal use in the Soviet Union and in East European countries is likely to increase to 1,500–2,000 mtce by the year 2000. The WOCOL study for Poland estimated that its domestic coal use will increase from 127 mtce in 1977 to 263 mtce in 2000.

Total World Coal Use

Combining the OECD projections with the figures for other areas indicates an increase in total world coal use from 2.5 billion tce in 1977 to 6–7 billion tce by the year 2000. This is an annual growth rate of 4–4.5 percent. Coal was in fact growing at about this rate during the 1950s but fell back to 1 percent per year during the 1960s and early 1970s, when competition with cheap oil was at its greatest.

---

4. Includes South Korea, Taiwan, the Philippines, Hong Kong, Malaysia, Thailand, and Singapore.

World Coal Import Requirements

Even though the world's present consumption of coal is large, most coal is consumed in the country where it is mined. International trade in coal in 1977 was only about 200 mtce (3 mbdoe), or 8 percent of total world coal use. Further, most of this trade was in high-quality metallurgical coals. International trade in steam coal was only about 60 mtce (less than 1 mbdoe), and most of this consisted of short-distance movements such as from Poland to the Soviet Union and Western Europe, and from the United States to Canada.

International trade in coal increased from 100 mtce in 1960 to 200 mtce in 1977. Most of the increase consisted of metallurgical coal shipments to Japan and Europe. By contrast the steam coal trade during the same period has remained approximately constant. Steam coal imports to the OECD countries actually declined between 1960 and 1973 before beginning to increase slowly again in response to sharply rising world oil prices.

Figure 2-10 shows our projected range of world metallurgical and steam coal import requirements to the end of the century. World coal trade increases by 3 to 5 times to 560–980 mtce. The higher

Figure 2-10   World Coal Import Requirements (1960–2000)

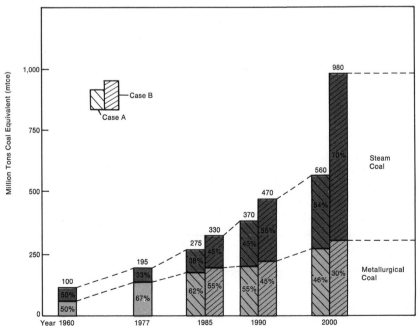

106

level is equivalent to about 13 mbdoe, or nearly half the oil exported from the OPEC countries in 1978. Within the total, the projected needs for imported steam coal increase by at least 5 times in Case A and up to about 12 times in Case B over the next two decades. As can be seen, steam coal as a percentage of world coal trade increases from 33 percent in 1977 to between 50 and 70 percent by the year 2000.

Steam Coal Imports

World steam coal import requirements in 1985 and in 2000 are shown in Table 2-4. Large increases are indicated for many na-

Table 2-4   World Steam Coal Imports by Country and Region (mtce)

Country/Region	1977	1985		2000	
		Case A	Case B	Case A	Case B
Denmark	4.6	10.7	11.1	9.4	20.9
Finland	4.1	3.4	3.4	7.7	12.4
France	14	11	34	26	100
Federal Republic of Germany	3ᶜ	9	11	20ᵈ	40ᵈ
Italy	2.0	10.3	10.9	16.5	45.5
Netherlands	1.5	7.0	7.0	19.9	34.2
Sweden	0.3	2.9	3.2	14.3	23.1
United Kingdom	1	—	—	—	15
Other Western Europe	7	13	13	32	42
OECD Europe	37	67	94	146	333
Canada	6	6	5ᵇ	8	4ᵇ
Japan	2	6	7	53	121
Total OECDª	45	80	105	210	460
East and Other Asia	—	5	24	60	179
Africa and Latin America	1	3	3	6	10
Centrally Planned Economies	17	20	20	30	30
Total Worldª	60	105	150	300	680

ª Totals are rounded.
ᵇ Under conditions of high coal increase in case B, Canada meets a greater share of its coal needs from domestic resources.
ᶜ Net imports, Federal Republic of Germany imported 8 mtce steam coal and exported 5 mtce in 1977.
ᵈ If indigenous production does not increase to the level of potential production estimated, FRG import requirements will be 20 mtce higher in the year 2000.

tions. Japan, which now imports only 2 mtce of steam coal, is estimated to require 25–50 times as much by the end of the century and to become the world's largest steam coal importer. Very large increases are also expected in some West European countries, for example France and Italy, and in the developing countries of East Asia, especially South Korea and Taiwan.

Metallurgical Coal Imports

Table 2-5 summarizes the projections for world metallurgical coal imports. The growing steel industries of the newly industrializing countries of Asia and Latin America account for about two-thirds of the expected increase in world metallurgical coal trade.

Table 2-5 World Metallurgical Coal Imports by Country and Region (mtce)

Country/Region	1977	1985		2000	
		Case A	Case B	Case A	Case B
Finland	0.9	1	1	1	1
France	10	11	12	12	15
Federal Republic of Germany	1	—	—	—	—
Italy	11.1	11.0	11.0	12	12
Netherlands	3.0	3.4	3.4	2.9	4.0
Sweden	1.8	2.2	2.2	2.8	2.8
United Kingdom	1	2	2	2	2
Other Western Europe	6	10	13	24	32
OECD Europe	35	40	45	57	69
Canada	7	7	5[b]	9	5[b]
Japan	60	73	76	79	85
Total OECD[a]	100	120	125	145	160
East and Other Asia	3	10	16	40	48
Africa and Latin America	7	20	20	57	80
Centrally Planned Economies	18	20	20	20	20
Total World[a]	130	170	180	260	300

[a] Totals are rounded.
[b] Under conditions of high coal increase in Case B, Canada meets a greater share of its coal needs from domestic resources.

Effect of Limitations on Oil Imports and Nuclear Energy Delays on OECD Coal Import Requirements

Figure 2-11 shows the possible effect of limitations on oil supply (Case A-1) and delays in nuclear power expansion (Case A-2)

108

on the steam coal import requirements of the OECD region. A 20 percent reduction in oil use projected for the year 2000 more than doubles the steam coal import needs in the Case A estimate, from 210 mtce to 460 mtce. This is shown by the Case A-1 import requirement. Another increment of 190 mtce will be needed if coal is substituted directly for a 30 percent reduction in Case A projections of OECD nuclear power expansion, in addition to the oil limitation. OECD steam coal imports needs would then rise to 650 mtce/yr (Case A-2 in Figure 2-11), which is considerably higher than the original Case B estimate and a 15-fold growth over the next 20 years.

Figure 2-11   The Effects of Oil Limitations and Nuclear Delays on OECD Steam Coal Import Requirements (1960–2000)

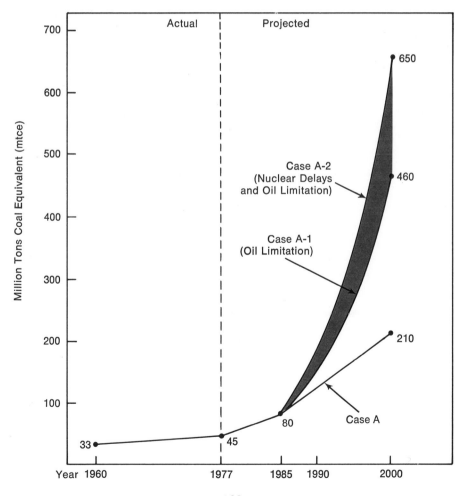

The projected evolution of coal imports is similar to the historic expansion of oil imports. OECD oil imports grew 10-fold from 1950 to 1970. In the 1950s oil imports grew from 2 to 8 mbd, which compares with the 1990s quadrupling pictured for steam coal imports, to 460–650 mtce/yr (6–9 mbdoe).

World Coal Export Potentials

The potential of coal as an international energy source depends on the ability and willingness of those countries with large coal resources to produce, transport, and export their coal in quantities sufficient to meet rapidly growing coal import requirements such as those projected in this study. Figure 2-12 presents the estimated coal

Figure 2-12   Coal Exporter Potentials—Year 2000 (mtce)

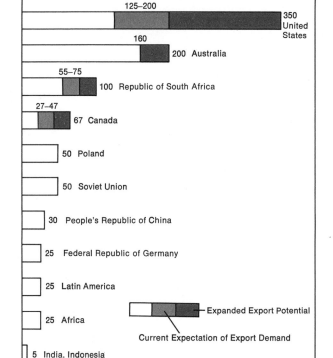

export potentials for the year 2000 taken from the country and regional WOCOL studies. The shaded areas for some exporters represent export levels beyond current expectations of demand. The expanded exports were specified by the country teams as feasible under favorable conditions if the demand develops soon enough.

Only Australia (160–200 mtce/yr) and the United States (125–350 mtce/yr) appear to be in a position to export significantly above 100 mtce/yr during this century. The other significant exporters by the year 2000 are projected to be the Republic of South Africa (55–100 mtce), Canada (27–67 mtce), Poland (50 mtce), the Soviet Union (50 mtce), and the People's Republic of China (30 mtce).

The growth of steam coal exports is projected to be particularly large. For example, Australia, which exported about 4 mtce steam coal in 1977, projects a potential of at least 85 mtce in the year 2000. For Canada, the corresponding figures are 1 mtce/yr in 1977 and a projection of up to 44 mtce/yr in 2000. The United States exported less than 5 mtce steam coal overseas in 1977, compared to projected steam coal exports of 65–280 mtce in the year 2000.

The Soviet Union and the People's Republic of China have large coal reserves that might be expected to allow them to increase their export levels above the values indicated in Figure 2-12, which are World Energy Conference estimates, if the demand develops and prices are high enough. There are major problems, however. Much of the Soviet reserves are low-calorific brown coal located in Siberia. Costly transport infrastructure requirements make it less likely that these reserves would be available for major exports as solid coal over the next 20 years.

The People's Republic of China also has large coal reserves that could support exports much higher than the WEC estimate of 30 mtce by 2000. However, the current policy appears to be geared to emphasize coal expansion for domestic use, relying on oil, the greater revenue earner, as the major energy export. As the People's Republic of China WOCOL report in Volume 2 states, "By the year 2000 our demand for coal will reach approximately 2 billion metric tons. The increases in China's coal production are mainly to meet domestic demand, with a small amount for export. However, since China has abundant coal reserves, the potential for coal export is considerably greater seen from a long-term point of view."

Some expansion in exports might be expected from developing

countries, Colombia in particular, but it is unlikely to be very large by the year 2000 on account of long lead times necessary for the development of coal mines, of the infrastructure necessary to move the coal to ports, and of the ports themselves. In addition rapid growth in domestic use and, in some cases, the quality of the coal are likely to limit coal exports from developing countries. Among WOCOL participants the situation is illustrated by India and Indonesia, which together estimate less than 5 mtce exports in 2000. This compares with an overall production level that grows from 72 to 285 mtce/yr between 1977 and 2000 in India, and from negligible levels to 20 mtce/yr for Indonesia.

Overall, our analysis indicates that the bulk of the growth in exports during this century will be provided by four countries: the United States, Australia, the Republic of South Africa, and Canada. Of the estimated world coal export potential in Figure 2-12, 75 percent lies in these four countries. Poland and the Soviet Union account for another 10 percent; however, the projections represent only a small increase beyond current export levels for these two countries. That the supplies of coal for international trade to the year 2000 will be dominated by the developed countries is a major contrast to world oil trade, which is dominated instead by the OPEC developing countries.

International Coal Trade Patterns

The largest coal flows in an expanded world coal trade system will be from Australia, Canada, Poland, the Republic of South Africa, and the United States to Japan and other East Asian ports and to Western Europe. The trading implications for exporters and importers are substantial. For example, if the landed (CIF) value of steam coal were $50 per tce, the estimated value of trade of 600–1,000 mtce/yr would amount to $30–50 billion annually. This would have significant consequences for balance-of-payments relationships.

The assessments by WOCOL teams from coal-importing countries included preferences for sources of coal imports as well as total coal import needs. These hypothetical preferences, with supplementary estimates for the regions not directly participating in the study, have been aggregated to provide a global picture. Table 2-6 summarizes the results.

Table 2-6 Hypothetical Trading Preferences for Sources of Coal Supply—Year 2000 (mtce)

Importer Country/Region	Case A — Exporter Source Country								Case B — Exporter Source Country							
	Australia	United States	South Africa	Poland	Canada	People's Republic of China	Other	Total Sources	Australia	United States	South Africa	Poland	Canada	People's Republic of China	Other	Total Sources
Denmark	1	—	2	3	1	—	2	9	2	2	6	5	2	1	3	21
Finland	—	1	—	5	1	—	2	9	1	2	—	5	2	1	2	13
France	7	4	8	5	3	1	10	38	28	24	20	7	19	4	16	115
Germany, Fed. Rep.	6	5	3	2	1	2	1	20	12	10	5	4	2	3	3	40
Italy	7	6	4	4	—	1	6	28	16	9	12	8	—	—	10	58
Netherlands	6	2	6	6	1	—	2	23	10	5	10	11	1	1	1	38
Sweden	4	2	—	4	3	—	4	17	7	3	—	6	5	—	4	26
United Kingdom	1	—	—	—	—	—	1	2	12	—	—	5	—	—	—	17
Other Western Europe	11	8	9	12	3	—	12	56	18	13	12	11	6	2	12	74
OECD Europe	43	28	32	41	13	4	40	200 [a]	106	68	65	62	37	13	51	400 [a]
Canada	—	17	—	—	—	—	—	17	—	9	—	—	—	—	—	9
Japan	45	43	5	1	18	10	10	132	78	66	6	1	25	16	14	206
East and Other Asia	34	23	15	—	20	4	4	100	86	52	33	—	48	—	8	227
Africa	—	1	—	—	—	—	1	3	12	10	—	—	5	—	3	30
Latin America	18	13	—	6	6	—	17	60	18	13	—	6	6	—	17	60
Centrally Planned Economies	—	—	—	25	—	—	25	50	—	—	—	25	—	—	25	50
Total World [a]	140	125	55	70	55	20	95	560	300	215	105	95	120	30	115	980

[a] Totals are rounded.

113

WOCOL teams preferred to diversify their sources of imported coal, even at the cost of paying somewhat higher prices for coal from some sources. Study of the table reveals that this diversification strategy is implemented differently in the European and Japanese projections.

The aggregated Western European trade flows indicate that Australia is the preferred source to supply 25 percent of the total European market. Roughly equal shares of 15–20 percent go to Poland, the Republic of South Africa, and the United States. Canada receives a 10 percent share, and the remaining 15 percent is scattered.

The Japanese projections, on the other hand, indicate roughly equal import shares for Australia and the United States, with each receiving one-third of the total market. Third and fourth are Canada and the People's Republic of China, each with a 10 percent share, with the remaining 10 percent scattered.

Such preferences reflect the current views of WOCOL teams of the attractiveness and availability of coal from the various suppliers, and take account of such factors as ocean freight costs and current coal prices. Such preferences might be expected to change over time to reflect changing economic and political conditions.

Feasibility of Importer-Exporter Trade Patterns

A review of the aggregated importer preferences in Table 2-6 for Case A (world total = 560 mtce in the year 2000) indicates little stress on export capacities of the three largest potential sources—Australia (140 mtce), United States (125 mtce), and the Republic of South Africa (55 mtce). European nations might have some difficulty obtaining their preferred amounts from Poland, but the excess is fairly small and could be made up elsewhere. At world coal trade levels of 500–600 mtce/yr coal importers should be able to exercise their preferred strategies of diversifying their sources of coal.

A different picture emerges under the conditions of the higher world coal import needs in Case B as well as in the A-1 and A-2 variations of Case A. This would put heavy pressures on the major coal exporters to expand exports beyond their current expectations (Figure 2-13). Further, most of the increase in requirements above 500 mtce/yr would likely have to come from two suppliers—Australia and the United States.

Figure 2-13 presents the comparison of current importer preferences and maximum export potentials as currently estimated for the 5 principal exporting countries; the excess requirements for 4 of these countries are substantial. On the other hand, the preference for imports from the United States—even though large at more than 200 mtce/yr—is considerably less than the 350-mtce/yr export potential estimated as technically and economically feasible in the U.S. assessment. Such export levels would amount to 15–20 percent of total U.S. production. Very large coal reserves are adequate to support such export levels as well as to meet domestic demand.

Figure 2-13 Importer Preferences vs. Potential of Key Exporters in the Year 2000

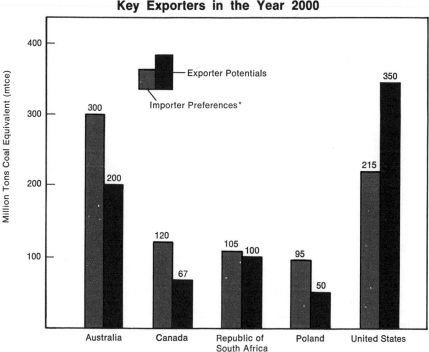

* Importer preferences are from Case B. Roughly similar requirements would be obtained from Case A-2, with Case A-1 needs about 20% lower.

If the projected world coal trade requirements are to be satisfied, the United States would have to be the largest coal exporter. Expanding U.S. exports to levels of 300–400 mtce/yr would require overcoming significant physical and institutional obstacles, including a general lack of awareness within the United States that there may be a demand for such expanded export levels. A recent government projection, for example, anticipated export requirements of only 100 mtce in the year 2000.[5] Moreover, the United States coal industry has been skeptical that a large export market will develop. Clearly, however, current expectations of potential exports on the part of the United States, and other countries, are subject to change as new markets develop around the world, as prices become more attractive, and as commitments by coal importers are demonstrated.

Implications for Trade Patterns

The general implications of this analysis of coal trade patterns are that at the high levels of world coal trade projected:

1. Coal importers, particularly in Europe, may need to accommodate their objectives of inexpensive sources of supply, diversification, and security to the realities of export capabilities of producing countries.

2. There will be heavy pressure on coal producers to export above their current expectations of demand. World coal trade levels of 800 to 1,000 mtce/yr by 2000 will be achievable only if the United States, Australia, and other countries with substantial coal resources greatly expand their coal exports.

3. Because of the long lead times involved, the high levels of coal trade projected will be realized by the year 2000 only if both producers and consumers are willing to make commitments in the early 1980's, even before all the uncertainties about future coal supply and demand are resolved.

National and Regional Implications of the Analysis

Virtually every country in the world will be affected by the rapid expansion of coal use and trade portrayed by our analysis. The major nations of Western Europe and Japan, which have relied on oil to provide most of the increase in their energy needs over the past two decades, must now make large investments to supplement and replace their oil-using infrastructures with coal-using equipment. Many of the world's developing nations face a similar prospect.

---

5. *National Energy Plan II*, U.S. Department of Energy (May 1979).

Coal-producing countries will need to respond quickly to growing demands for coal, both for domestic use and for export. Australia and the United States together may need to supply about half the export demand. There will be need for assurances that export customers receive as much consideration as domestic markets.

Ultimately it is up to both producers and consumers of coal to see that the required levels of coal use, production, and trade are developed. Exporters, particularly the large potential exporters such as the United States and Australia, will need to take positive steps to develop sufficient mining capacity, to provide adequate exporting infrastructure, to assure that their coal is competitively priced, and in general to relieve concerns of importers about their attractiveness as export sources. For their part, importers will need to identify their demand requirements and actively seek new trading partners.

Japan and the United States in particular will play pivotal roles in the energy transition we describe. Together they import today more than half of all the oil imported by the OECD nations. The United States especially has significant opportunities to reduce its oil import needs and use more coal. Our analysis shows that measures taken by these two countries to restrain oil import requirements will be central to maintaining a balance in world oil supply and demand throughout this century.

At the same time these two countries will be major driving forces in the expansion in the world trade of steam coal. Japan is likely to grow from a negligible importer today (2 mtce/yr) to probably the world's largest steam coal importer at 25–50 times that level, or more, depending on economic conditions in the next two decades. The United States will remain one of the world's largest coal exporters, and by the 1990s could become the world's balancing supplier of steam coal.

CHAPTER 3

COAL MARKETS AND PRICES

Coal Markets — Purchase Arrangements — Coal Costs — National Coal Policies — Observations

Coal Markets

Geographical Markets

Coal competes in two rather different types of markets—a geographic market and an end-use market. The price of a particular coal is generally determined by the interaction between the supply and the demand for that coal within the geographical area in which it can be sold on a competitive basis. The size of this geographical market area is determined by the delivered cost of competing coals, which in turn is determined by a combination of production costs (including labor and materials costs, depreciation, capital charges, royalties, and taxes), transportation costs, utilization costs (including pollution control), and the degree of market price competition. The boundaries of this market vary over time in response to shifting supply and demand conditions as well as to changes in government regulations and the prices of alternative fuels.

Because of low demand relative to production capacity, a widely distributed resource base, and high transport costs relative to price at the mine, steam coal markets have tended to be geographically limited. Metallurgical coal markets, on the other hand, have been more worldwide because metallurgical coal has a smaller, less widely distributed resource base, has almost no substitutes in use, and has traditionally received higher prices than steam coal.

As discussed in Chapter 2, over 90 percent of world coal production in 1977 was for end-use markets within the countries in which it was produced. Only a small portion moved in international trade, and most of that was metallurgical coal. WOCOL projects that by the year 2000, a larger percentage of world coal production will be internationally traded and that more of it will be steam coal. The United States, Australia, the Republic of South Africa, Canada, Poland, the Soviet Union, and the People's Republic of China are projected to account for virtually all of this export trade.

End-Use Markets

The principal markets for coal, as discussed in Chapter 2, are for generating heat in electric utilities, and for the manufacturing of coke for the iron and steel industries. Some coal is also used in other industries, and in the commercial and residential sectors. WOCOL projects expanding markets in all these sectors with the possible exception of residential and commercial use. Significant markets will be developing for coal in the industrial sector and for coal as a feedstock for synthetic fuels. Residential and commercial use could become larger in some countries through district heating plants and other coal conversion processes.

Coal Demand

The demand for metallurgical or coking coal depends directly on the demand for steel, although direct reduction and other technical developments are reducing the ratio of coke use to steel production.

The demand for steam coal is more broadly based and depends on a variety of factors including the level of economic activity in general as well as in specific economic sectors, on changes in the energy intensity of economic output, and on the cost and availability of alternative energy sources. Energy conservation—through better efficiency of use—will tend to reduce energy intensity, even though other factors, such as higher comfort heating standards, may tend to increase it.

Economic activity levels and energy intensity are both influenced, particularly in the longer term, by (1) the level of energy prices; (2) government actions (taxes, subsidies, regulations, etc.) that may or may not be initiated for energy policy reasons; and (3) continuing technological developments.

Coal's share of the total energy demand in any sector is limited by the types of fuel-burning equipment in use. Because most equipment has generally been designed to burn only one type of fuel, consideration of fuel alternatives will normally occur only to replace old plants with new ones, or to expand capacity. Adding a new plant entails a long lead time between the decision to add the plant and the actual completion of the installation. A new coal-fired electric power station, for example, takes about 4-5 years to construct, after the necessary permits and financing have been secured.

Changes in fuel-use technology will also have only limited effects on the demand for coal in the short term. For the immediate future the choice is among the existing types of energy-using equipment. New technologies for using coal, which are discussed in Chapter 7, will influence the demand for coal only over the longer term.

The choice of new plant will depend in part on the customer's economic assessment, at the time the choice is made, of the relative advantage of a given fuel over the life of the facility. Such conditions will of course tend to change during the 5–10 years before the plant is completed. Thus coal's near-term prospects are largely determined by past decisions made under past economic circumstances. Its intermediate future will be determined by today's assessment of that future and the investment decisions that result. Assessments of changes in technology and costs of transport and consumption, fuel availability, and prices of alternative fuels are several important elements in that investment decision process.

For example, in the economic evaluation between coal and nuclear power for electricity generation, utilities trade off the lower capital costs and higher fuel costs of a new coal-fired plant against the higher capital costs and lower fuel costs of a new nuclear plant. Judgments must also be made about the lead times required for nuclear stations; the future availability and price of uranium; technological considerations of safety, waste disposal, and pollution control; as well as the possible effects of public opposition to nuclear power.

Government policy can have a considerable effect on the choice of fuel. Legislation may, for example, restrict the use of nuclear power. It can also impose constraints on the production and use of coal. It can prohibit the use of fuels such as oil and natural gas in certain markets. National policy considerations such as the balance of payments, the need to provide employment, or security of supply may lead to policies that benefit an indigenous fuel over an imported fuel, even at higher cost.

The way in which policies are implemented varies from country to country because of differences in institutional systems. In some instances, especially in the power station market, decisions are concentrated in relatively few hands; in others they are dispersed over a large number of individuals as discussed in Chapter 8.

Purchase Arrangements

Long-Term Contracts

Coal is sold in a variety of ways. One common method is through a long-term contract between a coal supplier and a coal consumer that guarantees the consumer a reliable supply of coal of a specific quality within certain price ranges. The supplier benefits by securing collateral to help secure finance for new mines, by having a more predictable cash flow, and by avoiding the uncertainty associated with spot market sales. The consumer benefits by reducing price and supply uncertainties. The potential economic and political costs to an electric utility of not providing for a secure supply may be very high. Thus, utilities often agree to purchase their anticipated requirements from a given mine, or several mines, long before the mines are actually developed. In such cases the utility boilers are designed to match the characteristics of selected coals, although the more recent trend in many countries is toward building boilers flexible enough to use a wider range of coal types.

Long-term contracts may run 20 to 30 years. They usually include an initially agreed-upon base price, together with specified procedures by which the contract price is reviewed at regular intervals, or under unusual economic circumstances, to take account at least partially of escalation in costs of labor and materials, taxes, and other relevant items.

Recent contracts have recognized the need for greater flexibility in both price and quantity commitments. The cost-plus concept of some early contracts did not make adequate allowance for the effects of inflation over the life of the contract, or for the needs of the supplier to reinvest in further mine developments. Although the need for some flexibility in demand tonnage had been recognized, it was difficult to write contract clauses that would not severely disadvantage either buyer or seller. New long-term contracts are now sometimes written to provide for base price increases when quantity reductions exceed a certain limit. One of the problems of making contracts too flexible is the consequent reduction in the status of such contracts as security for project loans.

Spot and Short-Term Markets

Many steam coal users in the United States purchase 25 percent or more of their supplies on the spot market or on short-term (1-2 year) contracts, although some buy as much as 50 percent of their coal needs on a spot or short-term basis. As long as there is excess capacity they will be able to secure their coal requirements at or below contract prices. If, however, demand should exceed available supply, spot prices will rise above contract prices.

Other Arrangements

Alternative arrangements include captive mines, joint ventures, and contract mining. Some coal users (particularly steel companies) have purchased their own coal reserves and operate their own mines. Such a captive mine arrangement can offer increased certainty of supply and cost, but sometimes at a price higher than that possible from an independent supplier. In the United States in 1978 approximately 12 percent of the utility coal and 55 percent of the metallurgical coal was captive.

The consumer and the supplier both have an equity position in the mine in a joint venture arrangement. In contract mining the consumer owns the coal, but pays a mining company to mine it for a set fee per ton.

In an intermediary system an agent or broker acts on behalf of several mines and/or customers. This system could perhaps be expanded in the future to involve storage of coal.

Coal Costs

The development of a coal-mining project starts with exploration and the assessment of mineable reserves with regard to their extent, dimensions, geology, and quality, as described in Chapter 5. The potential producer will decide to exploit these reserves when he is satisfied that the project will prove profitable at the time that the coal will be produced, taking into account the expected coal demand, cost of production, availability of capital and manpower, and anticipated selling price. A long-term contract may provide sufficient financial assurances for mine development to proceed, though the investment will be subject to a variety of risks, including unanticipated inflation, cost overruns, exchange rate fluctuations, work stoppages, and even failure of consumer contract performance. In some cases, a coal producer may choose to assume greater financial risk and proceed to develop a mine without the assurance of a long-term contract.

Coal Production Costs

The cost of producing coal is a function of the costs of capital, material, equipment, and labor. These costs, in turn, are affected by the geological and environmental characteristics of the coal seams, including depth from surface, seam thickness, degree of faulting, nature of the roof or overburden, and whether surface mining or underground mining techniques are required. Other things equal, mining costs tend to increase as the seams become deeper or thinner.

Economies of scale theoretically permit a mine to be increased in size until minimum per unit cost is obtained. However, as a practical matter, this theoretically most efficient size is not often achieved because of other considerations, such as reserve size and contractual demands.

Although the vast subbituminous and brown coal deposits of the northern Great Plains of the United States, and some of the brown coal areas of the Federal Republic of Germany and Australia, are capable of supporting large mines of 20 million tons per year or more, in most coal fields the average mine size is much smaller. In the United States in 1977, for example, out of 6,200 mines only 44 mines (including 11 underground mines) or less than 1 percent of all mines produced more than 2 million tons of coal per year. Nevertheless, large mines are expected to play a greater role in the future, especially in the western United States, Australia, the Republic of South Africa, Canada, the Soviet Union, and the People's Republic of China.

In general, production costs are considerably lower for surface than for underground mines which are more labor intensive and which yield a relatively lower output per unit of labor. Surface mining, which is feasible only under certain geological conditions, usually uses a higher proportion of capital to labor, tends to achieve higher labor productivity, and also permits the recovery of a higher proportion of coal in the ground. Both surface and underground mining productivity varies considerably within and among coal-producing countries.

Land reclamation and other environmental protection costs have become an integral part of coal production costs. They can add from as little as $0.16 per ton (about 1 percent of total production cost) in the thick coal seams of the Powder River Basin in the western United States, to $3 per ton (about 10 percent of production cost) for the thin coal seams in surface mines in the eastern United States.

Other elements of coal production costs include royalties to coal deposit owners, other taxes, depreciation, and a rate of return on investment appropriate to the circumstances.

Estimating representative coal production and transport costs for an entire country has at best only very limited economic validity. For example the heat content of coals varies considerably. Lignite (brown coal) has about half the heat content of hard coal and it is much more expensive to transport on a heat content basis. Table 3-1 provides some indicative cost and price estimates from various coal sources to users in northwest Europe and Japan, based on summary data prepared for a recent U.S. Department of Energy optimum mine coal model. The actual cost of each component of each coal chain can change significantly depending on a variety of factors and the costs in the table should therefore be considered indicative only.

Coal Transport Costs

Transport costs for coal are generally higher, per unit of energy, than are those of oil, and the relative competitive positions of coal and oil vary according to transport distances. As with new mining capacity, the decision to invest in transport facilities also must be taken many years in advance of coal delivery.

Coal transportation costs, both inland and maritime, play a key role in determining the geographic extent of coal markets and coal's competition with other fuels. Transport costs may vary from very little in the case of users located near the mine to more than double the mine-mouth price for remote consumers, as illustrated in Table 3-1. Maritime shipping costs are discussed in detail in Chapter 6, and inland transport costs are discussed in each of the WOCOL country reports in Volume 2.

Within national markets, many coal consumers buy coal at the mine and then contract independently for its transportation by railroad, barge, or truck (freight on board or FOB the mine). The coal producer usually has little or no control over inland transport rates. In most parts of the world the railways are owned or regulated by the government. Hence, inland transport rates can be influenced by po-

Table 3-1 Indicative Steam Coal Costs and Prices
($ U.S. 1979 per metric ton)

	Price FOB Mine	Mine to Port	Price* FOB Port	Port Loading	Ocean Freight	Port Unloading	Delivered Price Range	Avg.	$/MBTU
To: NW Europe									
From: United States									
East—Underground	20-35	10-15	30-45	1-2	6-10	2	39-59	49	1.85
West—Surface	8-18	10-20	20-35	1-2	8-11	2	31-50	41	2.19
Canada									
West—Surface	15-20	10-20	25-35	1	8-12	2	36-50	42	1.92
Australia									
Underground	15-25	5-10	20-25	2	10-14	2	34-43	39	1.63
Surface	12-20	5-10	18-25	2	10-14	2	32-43	38	1.52
South Africa									
Underground	10-15	5-7	15-22	1	8-10	2	26-35	31	1.41
Poland									
Underground			23-31	1	5	2	31-39	35	1.46
To: Japan									
From: United States									
East—Underground	20-35	10-15	30-45	1-2	11-15	2	44-64	54	2.05
West—Surface	8-18	10-20	20-35	1-2	9-12	1	31-50	40	2.00
Canada									
West—Surface	15-20	10-20	25-35	1	8	1	35-45	40	2.00
Australia									
Underground	15-25	5-10	20-25	2	6-8	1	29-36	33	1.38
Surface	12-20	5-10	18-25	2	6-8	1	27-36	32	1.33
South Africa									
Underground	10-15	5-7	15-22	1	9	1	26-33	30	1.36
Poland									
Underground			23-31	1	11-13	1	36-44	40	1.67

* As mine prices and transport costs are given as ranges, FOB port prices are not necessarily the direct sum of the range limits.

Source: U.S. Department of Energy, *Coal Export Study* (1979), page 9, Table 4.

litical considerations and may not reflect true economic costs. In some areas in the United States, in particular, there is concern about the capacity of rail facilities to handle a greatly expanded coal production and export in the future, and there is also great concern among domestic customers and potential importers about the rapidly rising freight rates for hauling coal.

Coal Use Costs

In addition to delivered fuel prices, a fuel user must consider the costs associated with handling, storage, combustion, pollution control, and waste disposal. Depending on the coal sulfur content and heat content, as well as on air quality regulations, the costs of meeting environmental standards can add as much as $35 per ton of coal burned in some countries as described in Chapter 4.

The cost of coal-using facilities, particularly for small industrial consumers, tends to be higher than costs for oil and gas because of the handling charges. Hence, the delivered price of coal must be low enough to offset such disadvantages.

National Coal Policies

National coal policies and coal market operations differ considerably among countries. Some governments of the major consuming and producing countries tend to be more directly involved in market operation, than are others.

The coal industry in the United Kingdom is a statutory government undertaking run by the National Coal Board. The United Kingdom is virtually self-sufficient in coal, and both exports and imports a few million tons each year. Security of energy supply has been a major factor in determining national coal policy.

The Netherlands represents the other extreme; all its former coal mines have been closed, and there is a total reliance on imported coal.

In the Federal Republic of Germany the government has protected the indigenous coal industry since the late 1950s by import quotas, subsidies, and other measures in order to add security to the energy supply in sectors such as the steel industry and electricity generation, and to reduce the impact of a sharp coal production decrease on employment in the coal-mining regions.

127

The French government has a monopoly on coal production, and the government's coal-buying body, ATIC, is the single agency in France responsible for importing those quantities of foreign coal necessary to supplement domestic supplies. In Italy the government's electric utility, ENEL, presently imports all coal for electrical generation. In the longer term ENI, which has already started mining activity abroad, will become Italy's primary coal importer.

In Japan, the nine privately owned regional power companies and the Electric Power Development Company (EPDC), which has mixed private and public ownership, are likely to form a joint body for developing imported coal projects. Japan and South Korea both have small domestic coal-mining industries that receive some government assistance. Coal imports by Taiwan and South Korea are handled, in each case, by the electric power monopoly.

In Australia, coal mining operates within a federal-state regulatory framework. Though private companies produce coal for export and compete actively, the federal government must approve all coal export contracts. Development of Australia's extensive steam coal reserves to meet WOCOL export projections will require prices above 1979 levels, and may require some foreign capital. However, capital requirements should not pose serious impediments to coal development, and Australia's steam coal will continue to be competitive in many geographical markets.

In Canada, coal resources are owned and leased by the provinces in which they are located, but are produced by private companies except in Nova Scotia and New Brunswick, where Crown Corporations produce the bulk of the coal. The provincial governments are significant factors in coal development, and some provincial governments review coal contracts for acceptability of price.

In the United States, coal-mining companies are privately owned. However, most of the western coal reserves are owned by the federal government and are leased for development to private mining companies. Federal and state antitrust agencies have a mandate to ensure that viable competition is maintained in the coal industry, as well as in other nonregulated businesses or services. Federal and state governments also influence coal industry performance through environmental, health, and safety policies, through policies affecting coal transport and use, as well as through leasing provisions, taxes, and other requirements for coal development on federal and state lands.

Observations

In considering the formation of coal prices, it is necessary to distinguish between national and international markets. As described above, national markets in some countries are directly or indirectly influenced by government policies such as for import quotas, subsidies, and public ownership. In many countries, governments or public authorities own and produce coal, and use indigenous coal where possible for reasons of security of supply, maintenance of employment, or other national purposes. Most coal in the main coal-exporting countries, however, is privately produced and competitively sold. The newly emerging international steam coal market is therefore a relatively competitive system in which prices and quantities are determined primarily by market conditions.

In the steam coal market, the relevant price is the delivered price per unit of heat content, with adjustments for other quality parameters and for handling and utilization costs, relative to the delivered price of available alternatives.

Prices for a particular type of coal are determined by the interaction of the supply of that coal and the demand for it in the relevant geographical market, within a framework established by governments and taking account of competing coals and other fuels as well as conservation possibilities. Entry barriers, such as capital cost and economies of scale, have been low and many competitors have generally operated in each geographical market except those for which governments have limited entry.

The international steam coal market is just beginning to emerge and grow. This growth, coupled with relatively low entry barriers and the wide distribution of coal reserves around the world, suggests a vigorously competitive environment in international markets.

In order to make the full contribution that is necessary in the future, coal must remain economically competitive with alternative fuels, including unconventional oil and gas. Environmental and technical costs related to coal production, transport, conversion, and use may restrain some supply and demand quantities by raising production and consumption costs. However, even before the 1979 OPEC oil price increases, coal was competitive with other energy sources in many markets. The doubling of oil prices in 1979 has increased the economic attractiveness of coal substantially. In fact, despite large transport and environmental control costs, steam coal from Canada, the Republic of South Africa, Australia, and the United States has become economically attractive in the markets of many consumer countries. Further oil and gas price increases should continue to improve the competitive position of coal and increase its market penetration both geographically and in product markets.

Over the long term the real cost of coal is likely to increase, for a number of reasons: the development of new and more costly coal-mining facilities to replace or expand existing capacity; meeting new health and safety requirements; increasing expenditures on environmental protection and land reclamation; and meeting the rising costs of labor will all contribute to future cost increases. On the other hand rising productivity, greater mechanization, larger ships, and other economies of scale will tend to moderate such increases. These long-term coal production and transport costs will set the lower boundary to future coal prices. The maximum upper level will be determined by the price of the best available alternative.

Over the next two decades coal will primarily go into the heat and steam-raising markets. It is here that interfuel competition is strongest, although substitution is constrained by the user facilities and infrastructure in existence as a result of previous energy investment decisions. The outcome of the ongoing nuclear debate and the success, or otherwise, of the energy conservation programs now being proposed and implemented will also affect the demand for coal.

The short-term behavior of coal prices can, of course, differ substantially from long-term trends. If production capacity exceeds demand, as it does today, prices will tend to move toward the level of the variable cost of marginal mines rather than to be determined by the price of alternatives. On the other hand, if the quantity of coal demanded exceeds the supply, because of short-term supply and demand inelasticities, spot prices will tend to rise as producers and transporters attempt to capture the increased short-term revenue. However, such price increases would also provide economic signals to some coal suppliers to increase coal production, and thus would set forces in motion to eliminate, through competition, continuation of the short-term advantage.

Nonetheless, a large proportion of the future trade in coal will not be priced on a spot basis. It will be the subject of long-term contracts with greater price stability, and with prices more reflective of the cost of new as well as existing production and transport facilities.

In contrast to the world oil supply, which is becoming increasingly constrained, the coal supply has wide scope for expansion. Given sufficient lead time, it should be possible to increase supply to meet incremental demand at competitive prices. Moreover, the growing number of international coal suppliers, each with different interests, from all world regions including OECD, CPE and the developing countries, makes the formation of an international coal cartel unlikely.

In the long run, because of coal's abundance, and provided that a freely competitive international market is maintained, coal prices will most likely continue to be decoupled from those of oil. This will be particularly so as oil is increasingly removed from the heating market for use in transport and as a specialized petrochemical feedstock.

Government attitudes and policies will have a strong influence on the establishment and development of coal markets and will significantly affect the extent to which coal is able to fulfill its projected role in the future. Provided that adequate and stable economic and energy frameworks are available, capital and other resources should be available to meet the WOCOL projections. Governments can assist in the energy and coal market adjustment process by developing clearly defined energy and environmental objectives and by adopting a consistent, efficient set of energy policies to achieve those objectives. This would include removing or modifying obstacles to coal production, to international trade, and to coal use that are not clearly in long-term national interests. Failure of governments to provide such a framework will not only affect coal's ability to assist in meeting energy needs, but could entail substantial national and international consequences.

CHAPTER 4

ENVIRONMENT, HEALTH, AND SAFETY

Reasons for Environmental Concern — Surface Mining — Underground Mining — Occupational Health and Safety — Coal Preparation and Cleaning — Coal Transport and Storage — Coal Combustion — National Air Quality Standards — Emission Limitations and Strategies — CO_2 and Climate Change — Solid Waste — Thermal Emissions — Land and Water Use — Costs of Pollution Control for Utilities — Coal Use in Industries Other Than Utilities — New Coal Conversion Technologies — The Need for Research**

The large expansion of world coal production and use projected by the World Coal Study to the year 2000 means that each country will need to consider the resulting environmental, health, and safety issues. There is extensive experience with the mining, transportation, and use of coal and the application of environmental controls in the countries represented in this Study.[1] The major problems and issues to be considered in establishing environmental policies, standards, and laws have been identified. Research conducted over the last decade has improved the state of knowledge about both the issues and the control strategies and technologies available. By 1979 many countries had adopted detailed legislative and regulatory systems, or other less formal systems, for controlling the environmental, health, and safety effects accompanying increased coal production and use.

Reasons for Environmental Concern

Uncertainties remain about some issues. For example, the

---

1. In WOCOL Final Report Volume 2, *Future Coal Prospects: Country and Regional Assessments*, each of the WOCOL country teams describes the specific environmental, health, and safety standards now in effect in its country and provides indicative cost estimates for meeting these standards.

magnitude of long-term health effects of some of the emissions from coal combustion; the effects of fossil fuel combustion on global climate; and the environmental, health, and safety hazards posed by synthetic fuel plants have not yet been determined. There are trade-offs that must be made in each country among the degree of control, the resource and financial costs associated with that degree of control, and the benefits from using coal. Comparisons must also be made between coal and other energy alternatives that have environmental, health, and safety effects of their own. Although uncertainties make it difficult at this time to make universally accepted statements on environmental issues, four general observations can be made.

1. Most of the environmental risks from coal use are amenable to technological control. Emission release, noise, and other effects can be reduced to whatever level is required by applying currently available technology. Each increment of reduction increases the cost, and as one approaches total control, such costs become very large. Within what can be expected as standards we believe that coal can be produced, transported, and used cleanly at costs that leave coal competitive with other fuels. It is likely that environmental concerns or control costs will preclude the development of certain sites or certain coal resources. However, there are so many possible sites and resources remaining worldwide that such exclusions will not be a limiting factor to the expansion of coal use.

2. National perceptions of values differ on such things as exposure of the general public to health risks or visibility reduction in the atmosphere. For example, controversy continues on the extent of health effects from various emissions from coal combustion. Moreover, environmental impacts differ because of regional characteristics such as meteorology, topography, population density, and resource distribution. For such reasons, nations and regions take different positions on the kind and extent of environmental control measures they will require as coal use increases. Even though views differ widely, the countries in the World Coal Study plan large expansions of coal use and expect to apply measures that will ensure compliance with their national environmental standards.

3. There are some issues on which joint action among nations may be needed. Adequate mechanisms may not now exist for implementing international cooperation, although there are some precedents in the use of ocean resources and in the programs of OECD nations

on environmental matters. Agreement on the application of existing control technology for the interest of other nations in excess of what one nation might do solely on its own interest may be difficult but necessary. For example, the long-range transport of emissions and deposition of acid rain in several countries is receiving increased attention and may require early action. Similarly, improved understanding of the effects of pollutants requires continuing international cooperation. The need to integrate and coordinate some environmental actions at global, regional, national, and local levels is becoming more important.

4. Finally, there is concern about climate effects from the build-up of carbon dioxide (CO_2) in the atmosphere from combustion of all carbon fuels including oil, gas, coal, and wood. Currently there is uncertainty about CO_2 inputs from various sources, the absorption of CO_2 by various sinks, and the consequences of the effects of rising CO_2 content in the atmosphere. If the effects prove as serious as some researchers predict, the resulting situation would call for extraordinary kinds of international cooperation to control world fuel combustion or, alternatively, the amount of deforestation. Even though some people believe that immediate action is necessary, most expect that there are at least several decades to evaluate the CO_2 climate modification issue. We urge strong support of research to improve our understanding of the effects of CO_2 on climate and to expand studies of the impacts of climate change.

Figure 4-1 illustrates some environmental disturbances from coal-related activities. It shows only the by-products from these activities, not their possible impact on public health or on the ecosystem at large.

Surface Mining

Much of the increase in coal production to the year 2000 is expected to come from surface mines. Therefore satisfactory reclamation of land after mining becomes important in countries with large surface mineable reserves.

In some countries, such as the Federal Republic of Germany, reclamation has been practiced on a large scale for many years. Large sections of land are planned for mining many years in advance, towns are moved, and people relocated. After the land has been mined and

Figure 4-1 Environmental Disturbances from Coal-Related Activities

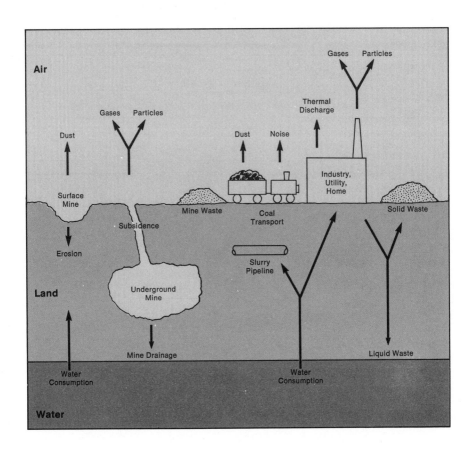

Adapted from *The Direct Use of Coal,* Prospects and Problems of Production and Combustion; Washington, D.C., Office of Technology Assessment, 1979, p. 184.

reclaimed, new towns are developed and people moved back into the area. Such large-scale activities require a clear commitment to the need; a suitable legislative framework; appropriate soil, climate, and depth of seams; long-range planning; and heavy investment for many years before the first coal is mined. For example, near Aachen, development of the Hambach deep-cast mine began in 1957, with a comprehensive drilling program. The first coal will be produced in 1982, and production will be increased until it reaches 50 million tons per year in the mid-1990s—40 years after initial drilling began.

In other countries, in an era of less concern for the environment, coal was surface mined and the landscape was destroyed. Even today in some regions, particularly in the developing countries, restoration is still minimal, especially if the costs are high. But most countries now have experience with land reclamation, and laws controlling it; "surface mining and reclamation" has become a single phrase. Moreover some mined-out areas are now being restored. For example, under current U.S. laws, a reclamation fee of $0.35 per ton of hard coal and $0.10 per ton of lignite from new surface-mined coal—a small fraction of the mining cost—is levied in order to finance reclamation of abandoned surface-mined land.

It is now possible with current technology to restore most surface-mined land to a condition equal to or better than the prior condition at a reasonable cost. Generally, reclamation is easier for flat areas. In arid regions, on steep mountain slopes, or in areas with fragile ecological systems reclamation is more difficult, and some such areas may not be licensed for mining under laws now in effect in several major coal-producing countries. Land areas critical for other purposes may also be excluded from mining. Australia, Canada, the Federal Republic of Germany, the United States, and the United Kingdom all have comprehensive legislation for the control and reclamation of surface-mined lands. The Surface Mining Control and Reclamation Act of 1977 in the United States is perhaps the most comprehensive legislation now in effect.

Illustrative cost estimates for surface mining reclamation are listed in Table 4-1. The actual costs per ton vary from mine to mine and according to the thickness of the seam being mined. The added cost per ton of coal to reclaim the land is in most cases not over 10 percent of its sales price; in many cases it is negligible.

Table 4-1 Illustrative Surface Mine Reclamation Costs (1977 U.S. $)

Area	Per Acre	Per Ton
U.S. Western coal (thick seam)	$3,000	$0.16
U.S. Central coal	5,000	0.89
U.S. Eastern coal (thin seam)	8,000	2.91
Western European average	5,000	Variable

Source: International Energy Agency, *Steam Coal Prospects to 2000* (Paris, Organization for Economic Co-operation and Development, 1978), p. 84.

Underground Mining

Underground mining, although less visible than surface mining, has its own set of environmental problems.

Underground mining can cause land subsidence—sinking into the area that has been mined. Where the room-and-pillar method of mining has been used, the problem of subsidence is usually not great. However, the room-and-pillar method leaves as much as half the coal in the ground. Another approach is to allow the surface of the land to subside after the removal of the underground coal and to provide for quickly and fairly carrying out repairs or giving compensation for any damage that may occur. This approach can be used only under certain geological conditions and usually when the long-wall method of mining is used. It is currently employed in the United Kingdom and other European countries. In the United States the cost of past damage caused by subsidence is being partially borne, under new legislation, by a reclamation fee on active underground mines of $0.15 per ton. In any case the cost of managing subsidence damage is a small percentage of the sales price of the coal.

Large quantities of solid waste are produced from surface and underground mining, as well as from coal preparation plants. This must be disposed of in surface piles that can be landscaped, as landfill, by returning to the mine, or by use as a construction material. In areas where the waste material contains contaminants such as high levels of sulfur, prevention of leaching requires careful control of water flows near the storage area, or may even require special ponding arrangements. Similarly, waste water pumped from some mines may contain such contaminants in amounts that must be controlled. However, in some areas this problem is not great. In fact the United Kingdom frequently puts such water to use in industry.

Occupational Health and Safety

Occupational health and safety are important concerns in coal mining. The major occupational health effect of coal mining has been the lung disease caused by breathing in the dust, pneumoconiosis (black lung disease). Reduction of dust levels by improved ventilation and filtration systems, dust suppression by water-spraying or by laying powdered limestone, and the application of strict work rules and practices have done much to reduce the risk of this disease. For example, the National Coal Board (NCB) of the United Kingdom reports that such measures have reduced by 40-fold in the last 20 years the incidence of pneumoconiosis in miners under the age of 35 years, as well as a similar reduction in other industrial diseases among miners. Improvements in occupational health are also reported by the mining industries of other industrialized countries.

Mine safety has similarly improved, if not quite as rapidly as improvements in health. In the United Kingdom, fatalities in deep mining are about 1 per million shifts worked. This record is 3 times better than those of the United States, the Federal Republic of Germany, France, or Belgium. It is more than a 5-fold improvement since 1952, which in turn was more than a 5-fold improvement on the rate 100 years earlier.

The causes of the catastrophic mine accidents of the past are now much better understood, and safety precautions are taken against the two worst hazards, gas explosions and flooding, both of which are now rare. Minor accidents have been reduced as well as a result of stricter legislation and regulation, the application of the results of research into safety, better training, and active safety programs. The technical improvements and greater mechanization of mining have also contributed to safety by reducing the number of miners per ton of coal produced. By designing out some of the hazards in the mining process, by removing men from hazardous areas, and by creating a high awareness of the need for stringent safety precautions, the mining industry has been steadily improving its health and safety performance.

In the United States, the new safety requirements of the Mining Enforcement and Safety Administration have considerably improved occupational safety conditions and have increased the labor cost of underground mining by approximately $4.00 per ton. With the stricter standards now enforced in most countries, occupational hazards of coal mining are becoming comparable to those of the construction industry or the high-risk manufacturing industries. The health and safety record in surface mines has already surpassed such a level. Still in some mining regions there is room for concern about the numbers of accidents that occur.

Coal Preparation and Cleaning

To date, coal-cleaning procedures have been designed mainly to remove some of the impurities in order to increase the heat content of the coal and to decrease the ash content being shipped and handled at the combustion site. Coal-cleaning procedures are being modified to reduce sulfur and trace elements in order to facilitate meeting environmental standards.

Mechanical cleaning processes based on differences in the specific gravity or surface characteristics can remove as much as half the sulfur content of some coals. Several chemical and mechanical processes are under development that may remove sulfur bonded chemically to the carbon in the coal.

Although coal preparation plants improve coal quality and thereby reduce emissions, they may themselves become significant sources of pollution. Up to 25 percent of the raw material mined, including some coal, must be disposed of as wastes. These wastes, like those from coal mining itself, have few uses and must be stored indefinitely and in a manner that minimizes the leaching of trace materials and soluble salts. Careful compaction and layering can reduce such pollution to levels that meet environmental standards.

Heat drying of the cleaned coal is expensive, uses energy, and may cause dust problems. The cost of meeting U.S. emission standards for particulates during coal drying is about $0.06–0.07 per ton of coal. Heat drying is now being replaced with mechanical dewatering, which costs less and avoids dust.

Recirculation and treatment of the wash water are integral parts of the operation of modern coal-cleaning plants, in order to reduce the amount of water used, eliminate discharge to streams, and allow recovery of coal fines. Compliance costs in the United States for waste water treatment are about $0.07 per ton of coal cleaned.

Such techniques are also used to clean up the acid water pumped from some mines.

Altogether the costs of meeting environmental standards in coal preparation are very small in relation to the sales price of the treated coal.

Coal Transport and Storage

Inland transport of coal is by truck or conveyor belt for short distances, and by barge or train for long distances. Transport across oceans is by typical bulk cargo-carrying ships. The principal environmental disturbances are dust, train noise, train or truck congestion, and the risk of accident causing property damage and risk to human life.

Dust can be controlled, by spraying with water and other techniques for approximately $0.05 per ton shipped. Oil is sometimes used and costs about $0.50 per ton of coal, including a credit for the heating value of oil added to the coal.

Coal slurry pipelines offer a promising alternative to barge and train for long-distance transport of coal. Located underground, they eliminate dust, railroad noise, and congestion, but require large quantities of water—one ton of water for each ton of coal transported. In some coal-mining areas water is in relatively short supply. This will either restrict the use of coal slurries, force the importation of water for this purpose, or require the use of some other liquid. Dewatering of the slurry at the receiving point and processing of the waste water can be done with available techniques for $0.15 to $0.25 per ton of coal shipped.

The effects of pollution from accidental spills during the transport are much less for coal than for oil. The risk and potential effects from accident, either in transport or in storage, for coal are not at all like those for liquefied natural gas.

Controlling dust problems and water pollution from leaching at coal storage piles at ports and at using facilities can be done at small incremental costs. Such coal piles must be managed, such as by compacting, to prevent spontaneous combustion from the reaction between the coal and atmospheric oxygen at ambient temperatures. This has been done successfully for many years. The visual impact of coal storage may, however, require that enclosed storage be used more extensively and that greater care be taken to protect the aesthetic and recreational value of adjoining areas. Moreover, land requirements for coal storage competes with the use of such land for other purposes.

Table 4-2 gives indicative costs for specific environmental protection measures in connection with coal mining, cleaning, and transport under conditions now prevailing in the United States. To get an indication of their scale, these costs can be compared with the delivered cost of steam coal in the United States which averaged about $25–30 per ton in 1977.

Coal Combustion

Coal combustion releases a number of different substances into the atmosphere. Greatly reducing the quantities of such products emitted into the atmosphere requires high costs for emission control. Cleaning up some of the emissions, especially sulfur, creates new waste disposal problems such as limestone sludge from flue gas desulfurization. Because there are substantial areas of disagreement among experts as to the effects of these emissions, it is not surprising that national policies differ widely on emission control goals and strategies.

Particles and gases from man-made sources together with dust particles and gases from natural sources are continually being released into the atmosphere, where secondary particles are formed by reactions among the primary particles and gases. Winds can transport these particles and gases for hundreds of miles, mixing them continuously. To determine the effect of one component of this mixture on the environment, or on the life expectancy of heterogeneous populations, is difficult at best. We do know that infrequent high concentrations of pollutants in the past, such as that which descended on London in 1952, can trigger a discernible increase in the death rate, especially among the elderly and chronically ill. Such incidents have occurred when meteorological conditions concentrated local emissions for at least several days.

Various gaseous and particulate substances from coal combustion at high concentrations are known to increase the rate of respiratory disease, aggravate asthma, cause headaches and chest pains, impair pulmonary functions, and cause general fatigue in susceptible members of exposed populations. Recent epidemiological studies do not provide clear evidence of a relationship between premature mortality and the sulfur oxide levels commonly found in the air of large cities. However, a slight correlation is observed between premature

Table 4-2 Indicative Cost Estimates for Specific Environmental Measures ($/ton of coal, 1977 U.S.)

	COAL MINING AND CLEANING				
	Contour Surface Mining (thin seams)	Area Surface Mining	All Surface Mining	Underground Mines	Comments
1. Reclamation of active mines (including prevention of mine subsidence)	2.80–3.00	0.15–0.90		1.00–5.00	Higher for surface mining in steep sloped areas
2. Fee for reclamation of abandoned mines			0.10 (Lignite) 0.35 (Hard Coal)	0.15	U.S. legislation
3. Dust control			0.10–0.20		
4. Mine drainage control	0.35–0.50	0.15–0.40		0.07–0.60	1985 technology
5. Occupational health and safety requirements				6.00	
6. Coal cleaning—prevention of runoff from storage and wastes			0.09	0.09	Per ton cleaned

	COAL TRANSPORTATION			Comments
	By Rail	Slurry Pipeline	Harbors	
1. Dust control, prevention of spills, control of runoff	0.05		Unknown	
2. Treatment of slurry water		0.15–0.25		Reduced by evaporating

Source: International Energy Agency, *Steam Coal Prospects to 2000* (Paris: Organization for Economic Co-Operation and Development, 1978), p. 93.

143

mortality and fine particulates that may come directly from particulate emissions or that may be created by the daughter products of the original sulfur dioxide or nitrogen oxide emissions. It is very difficult to distinguish effects of such products using correlations with such variables as socioeconomic class or prevailing weather conditions. Controlled human clinical studies to date show no significant discernible adverse health effects from "worst case" exposures to sulfates at levels several times the proposed U.S. ambient standards. The long-term effects of daily exposure to existing pollution levels, however, remain unknown. Nonetheless, many governments are adopting sulfur emission standards.

Emissions may also damage crops, fisheries, and materials. A wide variety of field, vegetable, fruit and nut, forage, and forest crops are sensitive to sulfur and nitrogen oxides under controlled exposures. Limited field studies to date indicate potential reductions in crop yield for some species, but increases in yields have been found in soils deficient in sulfates. Lakes in several parts of the world appear to have recently become acidic and, in a number of cases, the fish population has disappeared. Sulfur and nitrogen oxide emissions contribute to acid rain. It is still unclear what mitigation strategy would be most cost effective. As a remedy, some researchers have suggested the addition of lime to affected areas to buffer such acids.

Damage to nonliving materials from emissions is of concern in a number of countries. The soiling problem commonly associated with coal combustion has been all but eliminated, but remaining problems include deterioration of building materials and works of art, fading of dyes, weathering of textiles, and the corrosion of metals under long-term exposure to acidic deposition.

In addition, changes in visibility may occur in some areas. For example, some sulfur dioxide is converted in the atmosphere to sulfates that scatter light and may reduce visibility. This effect is most noticeable in dry regions such as the western United States, where the prevailing visibility may exceed 100 km. The problem of reduced visibility in cities has, however, been greatly diminished by the use of smokeless fuels or the virtual disappearance of small-scale residential and commercial uses of coal. The associated decrease in ground-level dust emissions has increased visibility. The elimination of the "pea-soup" fogs in London since the 1950s is an often-cited result of the requirement that only "smokeless" fuels be burned in the city.

National Air Quality Standards

Some countries control the potential adverse impacts of emissions on the environment by establishing national air quality standards that specify the maximum concentrations of certain chemicals permitted in the air. The principal standards, usually called primary air quality standards, specify the levels of pollutant concentration that cannot be exceeded in order to protect human health. Secondary air quality standards set limits on levels of pollutant concentration that cannot be exceeded in order to protect public welfare (vegetation, property, scenic value, etc.). Portions of present ambient air quality standards of several World Coal Study member countries are shown in Table 4-3.

Some countries do not have such ambient air quality standards. This does not mean that no attempt is made there to control pollution, but rather that different means of specifying and achieving air quality standards are used. Moreover, even in those countries with ambient air quality standards, methods of application may differ. In one case these standards may be stated as goals, to be achieved by whatever means are appropriate; in another they may include express prohibitions on the use of particular grades of coal or require specific actions to control emission sources.

Emission Limitations and Strategies

In order to achieve their stated air quality standards, national governments establish regulations limiting the rates of emissions from sources such as coal-fired electric power plants. In some countries, even more stringent standards are established by local governments. The major emissions that are regulated include sulfur dioxide (SO_2), particulate matter (total suspended particulates or TSP), and nitrogen dioxide (NO_2). Present national and regional standards for these emissions from new sources for several countries participating in the World Coal Study are given in Table 4-4.

Various approaches are used to determine allowable rates of emissions. Virtually all countries require some control of at least the larger particulates at the point of combustion. Some countries, such as France, the Federal Republic of Germany, and Italy, control the SO_2 concentrations by limiting the sulfur content of the coal burned. Others, such as the United Kingdom, rely on mechanical dispersion of emissions by tall stacks and prevailing winds. Intermittent control

Table 4-3 Illustrative National Ambient Air Quality Standards (mg/m³)

Country	SO_2	TSP	NO_2	NO	CO
Australia	No national ambient standard	No national ambient standard	No national ambient standard	No national ambient standard	No national ambient standard
Denmark	0.75[a]	0.25[a]	No national ambient standard	No national ambient standard	No national ambient standard
Federal Republic of Germany	0.14[b] 0.40[c]	0.2[b] 0.4[c]	0.1[b] 0.3[c]	0.2[b] 0.6[c]	10.0[b] 30.0[c]
Italy	0.25[m] 0.10[f]	No national ambient standard	No national ambient standard	No national ambient standard	No national ambient standard
Japan	0.14[d]	No national ambient standard	0.4[d] 0.8–0.12[e]	No national ambient standard	No national ambient standard
Netherlands	0.075[h] 0.25[c]	0.03[b] 0.12[c]	No national ambient standard	No national ambient standard	No national ambient standard
Poland	0.075[f] 0.35[g]	0.075[f, h] 0.2[g, h]	0.05[f] 0.2[g]	No national ambient standard	0.5[f]
United Kingdom	No national ambient standard	No national ambient standard	No national ambient standard	No national ambient standard	No national ambient standard
United States	0.36[d, i] 1.3[e, i]	0.26[d, i] 0.15[e, i]	0.1[k]	No national ambient standard	10.0[l]

[a] monthly average;
[b] long term;
[c] short term;
[d] primary standards (protective of human health);
[e] secondary standards (protective of public welfare, i.e., materials, flora and fauna);
[f] daily average for sensitive areas;
[g] daily average for non-industrial areas;
[h] particles less than 20μm;
[i] daily average;
[j] 3-hour average;
[k] annual average;
[l] 8-hour average;
[m] 30-min average.

Source: WOCOL Country Team Reports.

Table 4-4   New Source Performance Standards for Coal-Fired Power Plants (mg/m³)

Country	SO₂	TSP	NOₓ	CO
Australia	No standard	250	2,500	500
Denmark	No standard	150ᵃ	No standard	No standard
Federal Republic of Germany	2,845ᵇ	100ᶜ 150ᵈ	State of the art considered	250
Italy	2,000	No standard	No standard	No standard
Japan	500ᵉ 2,500ᶠ	200ᵉ 400ᶠ	767	No standard
Netherlands	No standard	No standard	No standard	No standard
Poland	No standard	No standard	No standard	No standard
United Kingdom	No standard	115	No standard	No standard
United States	1,900ᵍ	45ʰ	950ⁱ	No standard

ᵃ mg/Nm³
ᵇ converted from 2.75 g/kWh
ᶜ lignite
ᵈ hard coal
ᵉ urban

ᶠ rural
ᵍ converted from 1.2 lbs/10⁶ BTU
ʰ converted from 0.03 lbs/10⁶ BTU
ⁱ converted from 0.6 lbs/10⁶ BTU

Source: WOCOL Country Team Reports.

strategies are allowed in some countries such as Denmark—high-sulfur fuels may be used under favorable wind and weather conditions, and low-sulfur fuels must be used under adverse conditions. Some countries, such as the United States and Japan, rely on combined chemical and mechanical systems as well as on low-sulfur fuels to reduce emissions.

CO_2 and Climate Change

Because technical solutions for controlling CO_2 emissions are prohibitively expensive, and because large increases in the amount of atmospheric CO_2 may alter global climate, CO_2 emission poses one of the most perplexing problems resulting from the increased use of fossil fuels including coal. CO_2 is a trace gas in the atmosphere. In spite of its relatively small concentration (330 ppm) it has an influence on atmospheric temperature. It is largely transparent to sunlight, but it absorbs the infrared radiation emitted from the earth's surface and reradiates part of it, thereby reducing the rate of surface cooling. Consequently, it is thought that an increase in atmospheric CO_2 will contribute to a rise in the earth's temperature that has become commonly known as the greenhouse effect. Such an increase in the earth's average temperature would probably modify climate patterns, benefiting some regions but possibly bringing disaster to others.

For many reasons the issue of climate modification caused by increasing CO_2 in the atmosphere is more complex than the other environmental problems caused by fossil fuel combustion. There is a disagreement among scientists about the magnitude and urgency of the problem and about the detailed interactions involved.

CO_2 is absorbed, stored, and exchanged by the world's oceans, forests, soils, and sedimentary rocks in complex ways. This process is shown in a schematic form in Figure 4-2. Major sources of CO_2 are respiration from animals and decay of vegetation as well as evaporation from the oceans. Absorption of CO_2 by photosynthesis of plants and dissolving in the ocean are the sources of removal of CO_2 from the atmosphere. The input from fossil fuel combustion is small compared with the other fluxes, whose magnitude is not well established. In addition, the global atmosphere has been cooling since 1940 after an 80-year warming cycle. One of the difficult problems is to distinguish fossil fuel combustion effects from massive natural cycles of climate change.

People have been adding CO_2 to the atmosphere at an increasing rate since earliest times by destruction of the natural vegetative cover, changes in land use, and since the industrial revolution by the burning of fossil fuels. CO_2 in the atmosphere has increased by about 15 percent during the last century and is now increasing at about 0.4 percent per year. The effects of this are not yet predictable; natural feedback mechanisms such as increased cloudiness may act to moderate the greenhouse effect but such cloud cover, despite its high reflectivity to solar radiation, may reduce surface cooling even more.

On an energy content basis, coal combustion releases 25 percent more CO_2 than oil and 75 percent more CO_2 than natural gas. Large increases in coal combustion will have an effect on the level of atmospheric CO_2, but whether this will be significant in comparison with other mechanisms at work in the earth's carbon cycle is uncertain. Moreover, even if scientists agreed about the exact effects of CO_2 on climate, we do not now have international political systems capable of acting to prevent any further increases by restricting global fossil fuel combustion or by reducing the rate of deforestation.

Figure 4-2 Global Carbon Balance Sheet

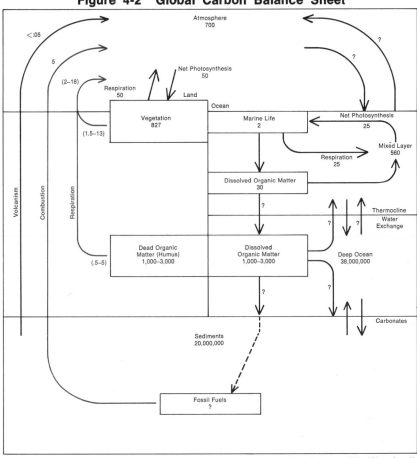

Source: From "The Carbon Dioxide Question," by George W. Woodwell. Copyright © 1977 by Scientific American, Inc. All rights reserved.

The Global Balance Sheet shows major carbon repositories and annual exchange rates among depots that are in contact. Quantities are expressed in units of 10^{15} grams, or billions of metric tons. Annual releases to the atmosphere governed by human activities are shown in color. Land plants fix a net of about 50×10^{15} grams of carbon per year. This carbon is either consumed and promptly respired by various terrestrial organisms or stored in the plant mass. The balance between fixation (net photosynthesis) and storage plus the total respiration of all terrestrial organisms determines whether there is a net flux of carbon dioxide to or from the biota. Many biologists now believe that there has been a long-term net flow of carbon dioxide from the biota into the atmosphere and that the flow is continuing. The carbon fixed by marine organisms is either respired or stored. It has been commonly assumed that most of it is respired immediately and recycled. It now seems possible that sinking fecal pellets may carry more carbon into the oceanic depths than had been thought. This transfer would supplement the normally slow diffusion of carbon dioxide into the surface layers of the ocean, where it comes into equilibrium with the carbonate-bicarbonate system. Although the deep ocean provides a virtually unlimited sink for carbon dioxide, gas must enter mixed layer and then penetrate thermocline, a thermally stratified layer that impedes mixing with deeper layers.

149

The issue of CO_2 climate modification requires sustained and expanded research efforts on both a national and an international scale. Progress in atmospheric theory is being made possible by improved models of global circulation supplied with much more extensive data. It may happen that some effects of CO_2 will become detectable on a regional and global scale before the end of the century, and will require a reassessment of world fossil fuel use at that time.

Solid Waste

The combustion of coal leaves behind a solid, unburnable residue called bottom ash. In addition, solid particles called fly ash are taken out of the flue gases by precipitators, filters, or scrubbers. Finally, the flue gas can be reacted with various agents to remove compounds from some of the gas, for example, lime or limestone slurries to remove SO_2. This produces a mixture of solids and water called sludge.

Some such solids are put to practical use. Fly ash and bottom ash are used commercially in cement making, in road building, as building materials, or as land fill. In the United States about one-third of the ash is so used, and in some European countries it is as much as one-half. Other avenues of use are being explored. The remaining materials must be disposed of in an acceptable way. Traditionally, both types of ash were disposed of by stacking in nearby land, frequently in lined pits to prevent water from leaching out contaminants. The cost of such disposal per ton of coal burned depends on land availability and lies in the range of $0.05 to $0.40 per ton of coal burned.

Sludges, on the other hand, are much more difficult to dispose of, and the cost for disposal is as much as $2 per ton of coal burned. Technologies are therefore being developed that result in a dry form of residual, such as gypsum, or in a useful product such as chemical sulfur.

Ash and sludge carry trace elements of materials contained in the original coal. Although some of these substances are toxic at high concentrations, it is unclear whether they are harmful when diluted as they are found in ash and sludge. A requirement that ash and sludge be disposed of as hazardous material could increase costs substantially. Depending on what needs to be done for disposal, the cost could reach $0.50 to $10 per ton of coal burned. In some countries, there is still much uncertainty about disposing of ash acceptably and what such processes would cost. In large-scale coal-consuming countries, nonconventional disposal centers may be needed in the future. Cost estimates for various waste disposal techniques are given in Table 4-5.

Thermal Emissions

Cooling is required in power plants and other large boilers whether fueled by coal, oil, gas, or uranium. Water is most often used for cooling. If the amount of heat is large and the water body receiving it is small, the thermal change in the natural environment can upset the ecosystem. To avoid such effects the heat can be dissipated by evaporation in a cooling tower at a cost of about the equivalent of $1 per ton of coal.

A great amount of the heat is dissipated by evaporation of the water, whether in a cooling tower or in a once-through cooling system. In areas where water is in short supply a completely closed system, much like an automobile radiator, can be used to dissipate the heat. Depending on the humidity and temperature, such systems cost the equivalent of $10 to $20 per ton of coal used.

Land and Water Use

The siting of coal production, transport, storage, and use facilities all involve the use of land. Such use competes with other functions for the land such as farming, residences, and recreation. In areas of dense population this competition is particularly acute. Methods of allocating the rights for use of such land differ among nations, but increasingly conflicting use is creating difficult problems to be solved by custom, marketplace, regulation, administrative fiat, or legal processes. In some countries and localities these increasing difficulties are leading to development of comprehensive planning for allocation of land and water among competing needs.

Similar conflicts over the use of water resources are also increasing. This involves the rights to use water for coal processes in mining, in slurry pipelines, in converting coal to gases or liquids, or for cooling in electric generating stations. Other uses of water may

Table 4-5 Comparative Cost Estimates for Specific Environmental Measures for Electric Utility Coal Utilization—New Sources (U.S. $/ton-1978)

	Australia	Denmark	Federal Republic of Germany	Japan	United States
Control of thermal discharges—cooling towers					
Wet	0.42	—	0.8–0.9	—	0.8–0.9
Dry	—	—	15.0–18.0	—	7.0–11.0
Particulate control					
ESP	0.80	1.00	2.0–3.0	1.6	1.5–2.0
Baghouse	—	—	—	5.4	1.6–1.8
SO_x Control					
Limestone FGD	—	—	20.0–25.0	14.2	7.0–18.0
Regenerative	—	—	—	—	17.0
Dry FGD	—	—	—	—	9.0–30.0
NO_x Control					
Combustion control (where possible)	0.20	—	—	0.8	0.2–0.5
Postcombustion control—NO_x selective	—	—	—	4.7	6.0–9.0
Postcombustion control—SO_x/NO_x scrubber	—	—	—	—	7.5–18.0
Combustion by-products disposal					
Ash—conventional	0.25	0.35	—	3.9	0.3–0.7
Ash—as hazardous material	—	0.45	—	—	5.0–10.0
Ash plus FGD sludge—conventional	—	—	—	—	2.5
Wastewater treatment					
Conventional	—	—	—	0.9	0.3–2.0
Zero discharge	—	—	—	—	2.0–3.0
Noise control—external plant only					
FD Fans	—	—	—	0.2	0.2–0.5

Source: WOCOL Country Team Reports. — = not available.

be agricultural, domestic, or industrial. Also the warmed water body may be viewed as an important fishing resource. These forms of competition for the water resource are leading to additional problems just as with the land resource.

Costs of Pollution Control for Utilities

The comparative costs for specific environmental measures for electric utility coal utilization in a number of countries are shown in Table 4-5. It should be stressed that environmental control measures usually require energy. For example, addition of sulfur removal facilities and a cooling tower may result in as much as a 10 percent loss of the output of a generating station. The energy costs are included in the estimated cost shown in Table 4-5.

Coal Use in Industries Other Than Utilities

Coal use other than for electricity generation is mainly in the metallurgical industry, where coal is used essentially as a chemical feedstock. Such coals are typically high quality, with a low content of sulfur and other impurities, and produce relatively low emissions of sulfur oxides and particulates. Such coals usually sell at a premium price. Their increased use, within the scale projected in WOCOL, may produce some environmental problems in some areas. Emissions from coke ovens represent special problems that can be handled adequately, although they may be expensive to deal with.

Steam coal, when it is used to generate electricity or raise steam in the industrial sector, may not have to meet the same environmental requirements as the same coal used in the utility sector. The same may be true for steam coal used in the commercial or residential sector to produce heat. On the one hand, because the amount of coal used at any location is so much smaller than the amount used by a utility, the environmental rules may be much less stringent. On the other hand, because such use is typically in areas of much greater concentrations of population and other environmental effluents, the environmental rules may be much stricter than for rural-sited generating stations. In any case, for identical emission standards, it is likely that the cost per unit of coal used will be high in small installations. Such a generalization, however, must be applied with caution because different technologies can be applied to achieve the control of emissions, and a technology applicable to only small-scale installations may be low in cost per unit of coal used. For example, if fluidized bed combustion is available and acceptable for small-scale installations, the cost of sulfur emission control per unit of coal could be considerably less than from a lime or limestone scrubber now used at generating stations.

New Coal Conversion Technologies

New coal conversion plants to make gases and liquids from coal will have to remove and dispose of ash, particulate matter, and sulfur and nitrogen compounds in the same way as do power plants. Some of these processes may have an advantage. For example, the ease of adding limestone to a fluidized bed to control sulfur may make it the preferred technology, but the amount of sulfur removed may not be sufficient to meet certain national standards.

About 20 percent or possibly more of the coal used by synthetic fuels conversion plants will be burned to supply heat and power for the synthetic process itself. Additional emissions will result from purification of the synthetic fuels. Depending on the technology used, these will take the form of potential air and water pollutants or solid waste, and they will all have to be controlled.

The extent of the environmental impact from coal-based synthetic fuels production is not clear. Synthetic fuel processes are complex; up to 80 classes of compounds of potentially hazardous substances may accompany the coal conversion process. In addition, there are currently very few data available on the characteristics of emissions from the various possible conversion processes because no commercial-scale plants have been built except in the Republic of South Africa. The major environmental problem will be controlling the production and release of potential carcinogens (primarily, complex organic compounds) during the coal conversion as well as controlling possible toxic materials in the waste. Because the basic costs of coal conversion will almost certainly be high, the additional costs of controlling emissions will probably be an acceptable fraction of the total cost given the market values and clean nature of the liquid and gas fuels produced.

The large quantities of water needed for all coal conversion processes may prove to be a major resource constraint on their development in areas in which water supplies are severely constrained.

Whether the various processes (high-calorific value gas, low-calorific value gas, solvent-refined coal, or liquefied coal) will represent a reduction of environmental pollution compared with the direct combustion of coal will depend on the degree of emission control used and the effectiveness of such control. Such evaluations are difficult at this stage of development of the different technologies, but it is likely that measures to control whatever environmental problems may arise will be found. Whether they will be cost effective is the question that must be determined by research.

The Need for Research

A number of improved technologies to reduce environmental effects from coal mining, transport, and use are currently under development. For example, improvements in underground mining to reduce the occupational hazards of miners, and better methods to remove sulfur oxide from flue gas, are well along in development. Such research is important to reduce the cost of environmental control and to improve the ability to remove contaminants from the environment. This includes the improvement in the workplace to reduce worker exposure to accident or health risks.

Recently, concern has been expressed as to the effect of small (submicron) particles on human health. Research is needed to determine whether the risks from such emissions are sufficiently high that further control is needed.

Also, a great deal of research has been under way in the last two decades to evaluate the effects of environmental contaminants. The research has involved primarily human health effects but also effects on ecosystems. It is this work that has made society more aware of the environmental risks it is taking, and has led to the environmental control strategies taken by various nations. It is important to continue such research so that additional controls can be aimed at those areas of greatest environmental risk, and so that control can be relaxed in areas found to be less necessary.

CHAPTER 5

COAL RESOURCES, RESERVES, AND PRODUCTION

Definition of Coal Resources and Reserves — Regional Distribution of Coal Resources and Reserves — World Coal Production — Coal-Mining Methods — Coal Preparation — Future Expansion Needs

The characteristics of coal vary widely, making precise comparison of coal resources in different locations difficult. The principal elements influencing coal quality are moisture, sulfur, ash, volatile matter, and fixed carbon. The calorific content of coal tends to increase with the presence of greater amounts of fixed carbon and lesser amounts of moisture and volatile matter.

Definition of Coal Resources and Reserves

Various systems of coal resources and reserves classification have been developed, and these systems differ from one country to another, and sometimes even among coal fields in the same country.

It is widely accepted that "geological resources" are a measure of the amount of coal in place, and "technically and economically recoverable reserves" are a measure of the quantity of those resources that can be exploited under current technical and economic conditions. There is no general agreement, however, on the precise manner in which coal resources and reserves should be differentiated. Moreover, statements of resources and reserves are not absolute but represent the state of knowledge and the conditions at the time of estimation and will change with new information resulting from further exploration or under different market conditions.

Estimating the total quantity of coal resources originally "in place" in known coal fields is relatively simple compared with estimating the resources of oil, natural gas, or uranium. This is because coal generally occurs in beds that tend to extend relatively continuously and laterally over considerable distances. With a knowledge of the general geological environment, the extent of a coal field can be estimated and used to derive the tonnage of coal resources in place.

Having made an assessment of the amount of coal resources in place, the next and more important step is to make an estimate of the amount of coal that is economically recoverable. In this regard it is also important to distinguish coal resources that are amenable to surface methods of extraction from those that can be mined only by underground techniques. The term "reserves" is generally used to designate those resources that are considered to be economically recoverable with modern methods. Because both economic and technical conditions differ regionally and also change with time, it is difficult to estimate how much of the resource base should be considered at any given time to be "reserves." Factors that influence this consideration include the coal's quality, the thickness of seams, the depth at which they occur, the extent and nature of the overburden, the location of the coal in relation to markets and transportation systems, and the availability and economics of competing energy sources. A coal deposit that in some regions of the world would be considered an economic source of energy, and therefore classified as reserves, may be uneconomic in other regions.

The World Energy Conference (WEC) has defined standards and established terminology for classifying coal deposits. The WEC classification divides coal into two major calorific categories: "hard" coal (including bituminous coal and anthracite) and "brown" coal (including subbituminous coal and lignite), based on the energy content of the coal. The dividing point is 23.76 megajoules per kilogram on a moisture and ash-free basis (which is equivalent to 5,700 kilocalories per kilogram or 10,250 Btu per pound). Coals with heat values above this cutoff point are considered by the WEC to be hard coals, whereas those with values below it are considered to be brown coals. It should be noted however that some countries, for example the United States and Canada, use the term "brown coal" to refer only to low-calorific lignites and not to the extensive resources of subbituminous coal that they possess.

The WEC differentiates hard and brown coals into geological resources and technically and economically recoverable reserves according to the seam depth and thickness characteristics given in Table 5-1. For example, hard coal deposits are classified as technically and economically recoverable reserves when the coal seam is thicker than 0.6 meters and is located less than 1,500 meters deep. Hard coals located between 1,500 and 2,000 meters deep are classed as resources regardless of seam thickness.

Table 5-1 WEC Specifications for Coal Resources and Reserves (meters)

	Geological Resources	Technically and Economically Recoverable Reserves
Hard Coal		
Minimum Thickness	Not Specified	0.6
Maximum Depth	2,000	1,500
Brown Coal		
Minimum Thickness	Not Specified	2.0
Maximum Depth	1,500	600

Regional Distribution of Coal Resources and Reserves

Coal deposits occur in a great many countries and in every continent. This distribution is illustrated in Figure 5-1, which shows the quantities of both geological resources and technically and economically recoverable reserves of coal by geographical regions for hard coal and brown coal, on a tons of coal equivalent (tce) basis. The figures are based on the 1978 WEC[1] work as updated in some cases by WOCOL country teams. It should be noted that not every country reports its coal resources on the basis developed by the WEC. In the Australian estimates for WOCOL, for example, reported hard coal resources include only coals to 1,000 meters depth and seams of greater than 1.5 meters in thickness except where accessible by surface mining, in which case seams thicker than 0.3 meters are included. For brown coal, Australia reports only resources to 200 meters depth in seams thicker than 15 meters.

Table 5-2 details world coal resource and reserve estimates by the main coal-producing countries.

---

1. Source: World Energy Conference, *Coal Resources*, (IPC Science and Technology Press, June, 1978.)

Figure 5-1   Geological Coal Resources of the World (10⁹ tce)

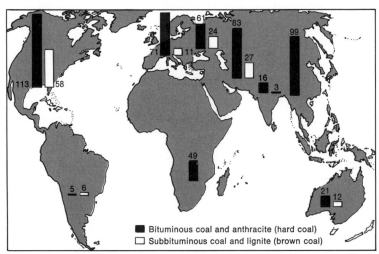

Source: 1978 WEC figures updated by WOCOL country teams. Non-WOCOL countries updated by Bergbau-Forschung.

Economically Recoverable Coal Reserves of the World (10⁹ tce)

Source: 1978 WEC figures updated by WOCOL country teams. Non-WOCOL countries updated by Bergbau-Forschung.

The magnitude of the world's coal resources is difficult to comprehend fully. The precise value of the resource estimate is in fact not of particular significance because the tonnages so estimated are so large in relation to current and future markets for coal. It may help to put the figures in perspective by noting that the current reserves alone amount to 660 billion tce or approximately 250 times the 1977

Table 5-2 World Coal Resources and Reserves, by Major Coal-Producing Countries (mtce)

	Geological Resources	Technically and Economically Recoverable Reserves
Australia[a]	600,000	32,800
Canada[a]	323,036	4,242
People's Republic of China[a]	1,438,045	98,883
Federal Republic of Germany[a]	246,800	34,419
India[a]	81,019	12,427
Poland[a]	139,750	59,600
Republic of South Africa	72,000	43,000
United Kingdom[a]	190,000	45,000
United States[a]	2,570,398	166,950
Soviet Union	4,860,000	109,900
Other Countries	229,164	55,711
Total World	10,750,212	662,932

[a] WOCOL member

Source: World Energy Conference and WOCOL Country Reports

world coal production. The known resources are 15 times higher. Moreover, the estimate of the world's reserves has increased by about 185 billion tce as a result of the greatly increased exploration since the oil price increases of 1973–1974. This increase in reserve estimates in the past few years is equal to about 70 years of production at the 1977 level.

Although there are quite significant coal resources in many parts of the world, Table 5-2 indicates that 10 countries account for about 98 percent of the currently estimated world coal resources and 90 percent of the reserves. Moreover, 4 countries—the Soviet Union, the United States, the People's Republic of China, and Australia—contain about 90 percent of the total resources and 60 percent of the reserves.

It is unlikely that major new coal basins will be discovered in countries with a long history of coal production. In many locations, however, exploration may be expected to delineate significant additional coal resources. In the developing countries, for example, exploration for coal has been much less widespread and less intense than exploration for oil, which was inexpensive, convenient to transport and use, and readily available.

Much of the present resource base is located in the northern temperate zone, where coal was used to fuel the early industrialization of many economies. Although the southern hemisphere is less favorable for coal deposits from a geological viewpoint (i.e., less extensive large sedimentary basins), there does appear to be some optimism that expanded exploration in the southern hemisphere and in the less developed regions of the northern hemisphere will result in the discovery of significant new coal reserves. For example, recent exploration in the southern part of Africa, particularly Botswana and Tanzania, as well as in Indonesia, is yielding favorable results. It is beginning to seem likely that the world's coal resources and reserves are significantly larger and more widely distributed geographically than was previously thought.

World Coal Production

Table 5-3 shows world production of coal by country and region in 1977 and projected in WOCOL Case B for the year 2000. The World Energy Conference, the World Bank, and other sources were consulted for the coal production estimates for countries and regions outside WOCOL. The 9 largest producers—the United States, the Soviet Union, the People's Republic of China, Poland, the Federal Republic of Germany, the United Kingdom, Australia, the Republic of South Africa, and India—account for about 85 percent of world coal production, both in 1977 and as projected in the year 2000. The largest three producers—the United States, the Soviet Union, and the People's Republic of China—account for about 60 percent of world coal production throughout the period.

As the table indicates, India is the only large coal producer among the developing countries. The World Bank in its recent publication *Coal Development Potential and Prospects in the Developing Countries* notes that coal output in the developing countries accounted for only about 5 percent of total 1977 world coal production. The report further pointed out, however, that about 50 developing countries have known coal resources and about 30 of these are currently producing coal. A large expansion in developing country coal production was projected. This is also reflected in the WOCOL coal projections, which show increases in production in India from 72 mtce in 1977 to 285 mtce in 2000; in Indonesia from less than 1 mtce in 1977 to 20 mtce in 2000; and from 25 to 180 mtce/yr over the same period for developing countries in Africa and Latin America. Strategies aimed at more rapid coal development in the developing countries must be

162

Table 5-3 World Coal Production by Country and Region (1977 and 2000)

	Production (mtce/yr)		% of World Total	
	1977	2000[a]	1977	2000[a]
Canada	23	159	0.9	2.3
United States	560	1,883	22.9	27.8
North America	583	2,042	23.8	30.1
Denmark	—	<1	<0.1	<0.1
Finland	—	—	—	—
France	21	10	0.9	<0.1
Federal Republic of Germany	120	150	4.9	2.2
Italy	<1	3	<0.1	<0.1
Netherlands	—	—	—	—
Sweden	—	—	—	—
United Kingdom	108	162	4.4	2.4
Other Western Europe	38	101	1.6	1.5
OECD Europe	287	426	11.7	6.3
Japan	18	18	0.7	0.3
Australia	76	326	3.1	4.8
Total OECD	964	2,813	39.3	41.5
Republic of South Africa	73	228	3.0	3.4
India	72	285	2.9	4.2
Indonesia	<1	20	<0.1	0.3
East and Other Asian countries	15	11	0.6	0.1
Africa and Latin America	25	180	1.0	2.7
People's Republic of China	373	1,450	15.2	21.4
Poland	167	313	6.8	4.6
Soviet Union	510	1,100	20.3	16.2
Other centrally planned economies	250	375	10.2	5.5
Total other regions	1,485	3,965	60.6	58.4
Total World[b]	2,450	6,780	100.0	100.0

[a] WOCOL Case B   [b] Totals Rounded
Source: World Energy Conference and WOCOL Country Reports.

designed both to increase production and to stimulate increased domestic use of coal.

Table 5-4 compares the projected cumulative coal production between 1977 and 2000 for WOCOL Case B with the currently estimated coal reserves for the major coal-producing countries and the world. It can be seen that even with the significant expansion of production that is projected, less than 16 percent of the world's currently estimated coal reserves would be consumed by the year 2000. This corresponds to only 1 percent of the presently known coal resources.

Table 5-4   Coal Reserves and Cumulative Production
(1977–2000) for Major Coal-Producing Countries (mtce)

	Technically and Economically Recoverable Reserves	Estimated Cumulative Production[b] 1977–2000	Cumulative Production[b] as a Percent of Reserves (%)
Australia[a]	32,800	4,200	13
Canada[a]	4,242	1,800	c
People's Republic of China[a]	98,883	20,000	20
Federal Republic of Germany[a]	34,419	3,100	9
India[a]	12,427	3,900	31
Poland[a]	59,600	6,700	11
Republic of South Africa	43,000	3,300	8
United Kingdom[a]	45,000	3,000	7
United States[a]	166,950	25,000	15
Soviet Union	109,900	18,000	16
Other Countries	55,711	14,000	25
Total World	662,932	103,000	16

[a] WOCOL member
[b] Based on WOCOL Case B
[c] The published estimates of Canadian reserves are not applicable for purposes of this comparison.

Coal-Mining Methods

Coal is mined by two main methods—surface mining and underground mining. The choice of methods is determined in large part by the geological parameters of the coal deposit. Only those coals that occur at relatively shallow depths are likely to be mined by surface methods in most countries. However, if the coal occurs in multiple seams where the overburden ratio is favorable, it is quite possible that relatively deep coals may be mined by surface methods. Quite different considerations of mining technology, environmental protection, health and safety, and costs apply to underground vs. surface mining.

Surface Mining

Surface mining is generally less costly and more flexible than underground mining, and recovers a higher proportion of the coal. It is characterized by higher labor productivity than in underground mines. The work force, although skilled, can be drawn from a number of other trades and related activities such as the metal-mining, construction, and transportation industries. There is often little need for special training.

164

There will therefore be a tendency to exploit the surface mine-able coals in preference to coals mineable only by underground methods. Surface mining is projected to continue to supply a large part of the coal produced in many countries, including the United States, Australia, Canada, and the Soviet Union. For example in the United States, an expansion of surface production from 375 million metric tons in 1977 to 850 mt/yr or more by the year 2000 is projected in the WOCOL country report—most of this projected expansion occurs in the western subbituminous coals.

The equipment used in surface mining—draglines, bucket wheel excavators, large trucks, high-capacity conveyors—has increased in unit size in recent years as a means of achieving economies of scale. However, recent trends indicate that the rate of increased productivity from the use of larger equipment is slowing.

There are a number of environmental issues associated with surface mining of coal including land use conflicts, land reclamation, disturbance of underground water, and potential pollution of surface waters. Technology exists to handle such problems in most locations at moderate cost. It is not felt that they will present insurmountable barriers to the increased production of coal from surface mines. Some coal, however, will not be mined because of government regulations designed to protect prime agricultural lands, urban areas, national parks, forests, and scenic areas. Environmental problems and methods of minimizing their impacts are more fully discussed in Chapter 4.

Underground Mining

The majority of the world's coal deposits are mineable only by underground methods. Even though surface mining is likely to expand considerably over the next few decades, it is probable that by the end of the century and thereafter the greater part of the world's coal production will come from underground mines.

Underground coal mining is usually more labor intensive than surface mining. Unit capital investment in new underground mines may in some cases be comparable to that for surface mines, but total costs per ton are generally higher in underground mines. Underground mines have less flexibility in operation than surface mines. The characteristic rhythm of production tends to be fairly uniform over the life of these mines because production capability is normally constrained by the designed haulage or hoisting capacity. The continuity of employment that accompanies stable production rates is very important.

A skilled, specially trained work force is required for underground mining—one that regards itself quite properly as a separate profession. Because coal mining has been carried out for decades or even centuries in some important old coal districts, coal mining inevitably has acquired many traditional aspects. It is not uncommon in these older production centers to have several generations of the same family all involved in coal mining. To the general public, however, such employment is not usually considered attractive; and in some countries it has been necessary to recruit both professionals and work force members from the less developed countries to adequately staff the industry.

Underground mining utilizes two main methods of extraction technology: room and pillar, and longwall. In room-and-pillar mining, the coal is mined at shallow working depths by cutting panels or "rooms" into the coal and leaving "pillars" of coal behind to support the roof of the mine. Because of the pillars, the total recovery of coal from the deposit is relatively low—on the order of 50–60 percent. Although in some cases the pillars are recovered after mining is completed, in most cases they are left to support the roof and thus prevent or lessen the potential for subsidence of the overlying material, including the surface land. If severe subsidence does occur it may cause damage to surface structures and disrupt agricultural lands overlying the coal mine. In most Western European coal mines the considerably greater working depth as compared with those in U.S. coal mines precludes use of room-and-pillar mining, mainly because of the high rock pressure encountered at these depths on the roofs of the "rooms."

The longwall mining method involves the use of shearer-loaders in conjunction with fully mechanized face support systems. Powered supports are used to temporarily support the roof while the coal is being removed. Because no pillars need to be left to support the roof, it is possible to recover a much higher percentage of the coal in the deposit. It is becoming more widely agreed that in terms of output capacity, the longwall operations are superior to the room-and-pillar workings. The cut per face per day produced by longwall mining may be up to 4 times higher than that achieved in room-and-pillar mining. The investments required for equipment for a modern longwall are not substantially higher than those required in room-and-pillar mining. After the coal is removed, the power supports are also removed and the roof is allowed to collapse. Because the entire roof is allowed to fall, there is a relatively uniform subsidence and thus the impact on the surface topography and structures is minimized. However, this method is suitable only for use where the roof material is such that it falls in a uniform and predictable manner after the removal of the power supports.

Because of increasing work depths in coal mines in the United States, room-and-pillar mining is being more and more affected by the adverse effects of rock pressure. The growing number of longwall faces in operation in the United States for example—about 20 percent of underground coal production in 1978—reflects this trend.

Existing underground mining technology is most effective when the coal seams are of reasonable thickness, nearly horizontal, with good roof and little or no gas, and occurring at moderate depths. As the mining conditions deviate from such optimum conditions, difficulties and costs increase accordingly. Productivity is being improved by the increasing automation of all aspects of the mining process. Increasing automation involves developing more reliable equipment, providing means to monitor the equipment, and developing methods of control. Computers and microprocessors are playing an increasing role. In the future, remote-controlled mining methods may prove applicable for recovery of the substantial resources of coal that occur in seams too thin or too deep for extraction by existing techniques.

It is not yet clear how soon such methods will prove practical for widespread application. Techniques have been developed through the science of rock mechanics that give better prediction of the roof and ground conditions likely to be encountered, especially at great depths or under disturbed conditions. Cooling and ventilation methods have been applied to cope with the high temperatures that occur at great depths. In all high-production technologies improved haulage methods are a necessity, and conveyors of greater flexibility have been devised. Also promising are coal-water slurry techniques to transport the mined coal from the face by hydraulic hoisting of coal to the surface. Improved shaft-sinking methods have reduced the time needed to open deep mines in some regions. In summary, there are many opportunities for research and development into mining technology to improve the safety and efficiency of deep-mine coal extraction.

Coal Preparation

Coal that moves in international trade is nearly always "prepared," whether destined for the energy market or the metallurgical market. Normally, the objective of the preparation process is to reduce the mineral matter and/or sulfur content of the raw coal and to improve the uniformity of the coal shipped.

Research is also underway to reduce the naturally high water content of low-rank coals; this, along with resolution of the problem of lignite susceptibility to spontaneous combustion, would allow such coals to be sold and used more competitively in more distant markets. Successful use of thermal drying and/or of briquetting of coals could be a key factor in determining whether low-rank coals can be exported and traded internationally.

Future Expansion Needs

The economically recoverable reserves of coal are very large—many times those of oil and gas—and sufficient to support the greatly expanded worldwide use of coal well into the next century. Although a large fraction of currently known coal reserves are located in relatively few countries, coal resources are known to exist in many countries in the world, and there are good prospects that these resources will be able to support coal use in a greater number of countries in the future. It will be necessary to develop many new coal deposits and open hundreds of new mines in order to realize the projected expansion in world coal production. Such actions require comparatively long lead times and must be initiated years before the coal is delivered. In addition, it will be necessary to construct extensive infrastructure facilities, including in some cases towns, railways, and ports. Major programs of recruiting and training of personnel, especially for underground mining, will be required. All activities must be carried out in ways that conform to environmental standards. The timely investment of capital will play a decisive role in the future availability of coal. Prompt and effective actions and investment decisions must be taken well in advance of the actual need to ensure that the mining capacity of the world coal industry is available when needed.

CHAPTER 6

MARITIME TRANSPORTATION AND PORTS

Coal Trade Routes — Shipping Requirements — Freight Rates — Bunkers (Fuel for Ships) — Coal Handling — Ports and Terminals — Port Requirements for Future Coal Trade

Ports and the ships that ply between them are vital links in the chain from the coal mine in the exporting country to the consumer in the importing country. The coal transport chain begins with the movement of coal from inland mines to ports where transshipment to ocean-going vessels takes place. From here the coal moves in international trade to tidewater ports in consuming countries. From such ports the coal has to be transferred for shipment to the ultimate consumers, who may be located inland. The likely routes and potential problems associated with the inland portions of the coal transport chain are discussed in the WOCOL country reports (see Volume 2). This chapter is designed to discuss the key factors and issues involved in the port-to-port part of the chain, that is, the maritime transport and ports.

With a 3–5-fold expansion in international coal trade, and with a 10–15-fold expansion in the international trade of steam coal in prospect, the question of ships and port facilities is obviously a crucial one. Transportation costs are an important element in the total delivered cost of coal and may be the deciding factor in establishing the balance of competition between coal and other energy sources. More fundamentally, unless the required new ships and ports are available in time, the projected expansion in international trade in coal would be delayed.

Coal Trade Routes

The most significant movements of coal will develop between

the major industrialized and industrializing countries with energy deficits, and those countries with large reserves of coal (Table 6-1).

Table 6-1 Maritime Coal Trade Routes
(nautical miles)

Source	Destination	Cape Route	Canal Route
Canada (West Coast)	West Europe (ARA)ᵃ	15,400	10,000
	Japan	4,800	—
United States (East Coast)	West Europe	3,600	—
	Japan	16,000	10,000
United States (West Coast)	West Europe	13,800	8,400
	Japan	4,750	—
Republic of South Africa	West Europe	7,200	—
	Japan	8,700	—
Australia	West Europe	13,700	11,400
	Japan	3,600	—

ᵃ ARA = Amsterdam, Rotterdam, Antwerp.

Because vessels larger than about 65,000 DWT[1] (Panamax) are not able today to pass through the Panama and the Suez canals, larger ships must use the Cape Horn or Cape of Good Hope routes. In some cases this leads to a considerable lengthening of a round trip. The principal maritime coal trade routes are shown in Figure 6–1.

Shipping Requirements

Coal is carried either in dry bulk vessels that are also capable of carrying other commodities such as iron ore, grain, bauxite, and phosphate rock, or in so-called combination carriers that can carry oil as well as dry bulk cargoes. Coal, iron ore, and grain are by far the most significant of these dry bulk commodities. In WOCOL Case B the volume of international coal trade in the year 2000 is projected to be almost 1,000 mtce/yr. Using simplifying assumptions of ten round-trips per year and no backhaul cargoes, approximately 1,000 ships of 100,000 DWT would be required for this trade.

The total active and inactive world dry bulk carrier fleet in 1979 amounted to about 135 milion DWT and the total active and inactive combination carrier fleet to about 48 million DWT. Historically the percentage of combination carriers trading in dry bulk has varied between 35 and 65 percent.

1. Deadweight tons. The deadweight is the loaded displacement minus the empty displacement and measures the cargo capacity of a vessel.

Figure 6-1 Maritime Coal Trade Routes

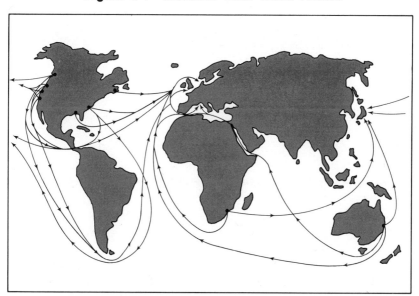

In order to estimate the total requirements for new ships and hence the pressure on world ship-building capacity, assumptions have to be made about the total increase in demand for all commodities, the life of ships, the rate of scrapping and loss, and the proportion available for dry bulk commodities. In addition, estimates need to be made about the requirements for new oil tankers. Because of the many technical and economic similarities between dry bulk carriers and oil tankers it is necessary to consider them together when assessing the building potential of the world's shipyards. The total bulk tonnage requirement (dry bulk carriers and oil tankers) determined under a representative set of assumptions grows to a peak averaging about 50 million DWT per year in the period 1985–1990 and declines thereafter.

The maximum output of the world ship-building industry was in 1975, when the amount of ships produced was 35.9 million GRT.[2] This peak was reached as a result of a continued build-up of output from a figure of about 8 million GRT in the beginning of the 1960s. The increase between 1970 and 1975 alone was almost 15 million GRT. After 1975, however, demand declined and output amounted to only 15.4 million GRT in 1978. Faced by this sharp decline in

---

2. Gross registered tons, the legal definition of the maximum cargo-carrying capacity of a ship, roughly equivalent to deadweight tons (DWT).

demand, most shipyards cut their work forces and some have been forced to cease operations entirely.

Nevertheless, the reductions that have been made are mostly based on layoffs and the real production potential (based on physical facilities and design and management teams) remains largely intact. Furthermore, expansion of ship-building capacity has continued in many countries.

It is therefore believed that the 1975 ship-building capacity or higher can be reached again, without much difficulty, given the demand and a reasonable lead time. The rate of increase demonstrated between 1970 and 1975 adds further support to the view that the world ship-building industry would have no difficulty in expanding to meet the requirements for new vessels to handle a rapidly expanding coal trade. In fact, the coal ship-building program—averaging about 50 coal ships or 5 million DWT per year for 20 years—would be of substantial economic benefit to the world ship-building industry.

Freight Rates

In general, the larger the ship the lower will be the unit operating costs and hence, the lower the freight rate. The present maximum size of ships operating in the dry bulk trade is about 150,000 DWT. During the 1980s the major part of coal in seaborne international trade is expected to be carried in vessels of about 100,000–125,000 DWT. In the 1990s, ships as large as 250,000 DWT may come into use. Ships are not expected to be built much in excess of 250,000 DWT because the small further gain in freight differentials is unlikely to be sufficient incentive for the extra investments required to install larger machinery onshore and to provide increased water depth.

In Table 6-2 some of the effects of ship size on freight rates can be seen. The rates used, known as equilibrium rates, are theoretical figures used for project evaluation proposals only, representing the long-term average freight rate that an owner would seek, over the lifetime of his ship, to give a reasonable rate of return on his capital investment. In Table 6-2 the equilibrium rates are shown for three ship sizes for various journeys. In each case it is assumed that ports capable of handling the vessel size exist at both source and destination.

It frequently happens, however, that a coal port is capable of accepting large ships in every particular except for draft (harbor depth), and in these circumstances it is often advantageous to operate

Table 6-2 "Equilibrium" Freight Rates for Selected Routes
(1979 U.S. $/metric ton)

Destination	Western Europe		
Source	Australia	W. Canada	South Africa
Single port			
150,000 DWT	13.8	14.1	10.7
115,000 DWT	14.9	14.9	11.7
75,000 DWT	17.6	14.1[a]	14.1
Two ports			
150,000 DWT and 75,000 DWT draft	14.8	15.1	11.4

[a] Via Panama Canal, which makes the smaller vessel equally attractive at current rates for canal dues.

the ship with two-port loading or discharging. The shallow-draft port is served by a large ship that has first partially unloaded its cargo in a deep-water port to allow it to meet the draft restriction. Table 6-2 demonstrates the advantage that this procedure brings in the case of a port that is limited to 75,000 DWT draft but that can otherwise accept the dimensions of a 150,000 DWT vessel.

A great many coal ports have neither the draft nor the other necessary dimensions to receive large bulk carriers and must be supplied by smaller vessels. In these circumstances, when the distances from the supply to the receiving port are large, transshipment may prove advantageous. For transshipment to be possible, a large deep-water port with both ship unloading and reloading facilities is required. The major part of the journey is then accomplished in a large vessel that discharges its cargo at the transshipment port. From here the coal is reloaded into a ship of suitable dimensions to enter into the ultimate receiving port. Table 6-3 provides some typical costs for this procedure.

Table 6-3 Effect on Shipping Costs of Transshipment
(1979 U.S. $/metric ton)

Destination	Western Europe		
Source	Australia	W. Canada	South Africa
150,000 DWT	13.8	14.1	10.7
Transshipment cost	4.0	4.0	4.0
25,000 DWT	7.0	7.0	7.0
Total cost	24.8	25.1	21.7
Direct by 25,000 DWT	32.0	28.0	25.0
Savings by transshipment	7.2	2.9	3.3

173

Another significant element in the determination of bulk freight costs is the extent to which a backhaul cargo is available to occupy the otherwise empty coal ship on its return to the coal-loading port. In the typical costs shown in Table 6-2, the South African freight rates are calculated on a round-trip basis without backhaul cargo (coal to Europe, empty back to South Africa) whereas for Canada and Australia the cost has been calculated by assuming a credit for backhaul cargo. For the 150,000 DWT vessels used in Table 6-2, the difference between a backhaul leg and a round-voyage basis would be approximately as shown in Table 6-4.

Table 6-4   Effect on Shipping Costs of Return Cargoes
(1979 U.S. $/metric ton)

Destination	Western Europe		
Source	Australia	W. Canada	South Africa
150,000 DWT Round voyage	20.1	19.0	10.7
150,000 DWT Backhaul assumed	13.8	14.1	5.8
Savings by backhaul	6.3	4.9	4.9

In the future the projected rapid growth in maritime coal transportation is unlikely to be matched by a similar growth in other commodities. Backhaul opportunities will therefore be limited, and consequently it is likely that coal will be carried in large ships specially designed for the coal trade.

Bunkers (Fuel for Ships)

A wide variety of fuels can be considered for use as bunkers, but for economic and technical reasons the choice is limited to only a few. Liquid hydrocarbon fuels are currently the most extensively used bunker fuels and range from diesel and gas oil to high-viscosity residual fuel oils. However, in a changing economic environment alternative fuels, especially coal, are likely to be considered in the future. Gaseous fuels are almost exclusively restricted to the use of cargo boil-off in LNG carriers. Nuclear-powered ships are utilized for military purposes but have had no applications in the merchant marine other than as a few subsidized demonstration vessels.

Current worldwide liquid bunker fuel requirements are 2 mbd and are expected to grow to about 4–5 mbdoe to meet the needs of the shipping fleet in the year 2000. This is equivalent to 15–20 percent of the oil currently exported by OPEC countries. The cost of marine fuels has followed the general increase of world oil prices that began in 1973–1974. By 1979 both marine fuel oil and marine diesel oil were nearly 15 times the level of 1970.

Fifty years ago the world's navies and merchant fleets were entirely fueled by coal, and coaling stations were located along all the trade routes. After 1950, however, oil replaced coal in hardly more than a ship generation. Present and expected future constraints on the availability and the price of oil are causing serious reexamination of the potential use of coal again as bunker fuel for ocean-going vessels. Coal-fired ships appear increasingly attractive economically, because of expected fuel cost savings, despite the fact that coal-fired ships cost more to build than oil-fueled ships, and coal is more bulky and difficult to store and handle than oil. Some Australian companies have completed feasibility studies and are inviting shipbuilders to bid for coal-fired bulk carriers, for transporting coal to Asian ports and minerals around the Australian coast.

By the year 2000 a substantial proportion of the world's merchant fleet may once again be coal fired. Initially, this would be expected for coal-carrying ships—the bunker fuel requirements for these ships alone could approximate 50–60 mtce/yr by the year 2000. With the build-up of coal trade and the establishment of coaling stations along trade routes, coal-firing for other parts of the world's merchant fleet could grow rapidly. Such a trend would be reinforced by the relatively rapid turnover of the shipping fleet—the lifetime of a ship is typically about 18 years, so that essentially the entire merchant fleet operating in the year 2000 would be built during the 1980–2000 period.

Coal Handling

Ship unloaders are usually of the conventional clamshell type. Their dimensions and capacity have increased tremendously with the growth in ship size, and machines are now in existence that can unload 2,000-3,000 metric tons of coal per hour from large carriers. However, average unloading rates are seldom higher than 1,000 to 1,500 t/hr, and several such machines are usually required to serve one vessel. In an attempt to increase the unloading rate and reduce the number of unloaders, several designs for continuous unloaders have been put forward, but so far only a few have been installed for smaller vessels.

The ship, barge, and coastal vessel-loading installations of transshipment terminals usually consist of a structure traveling on the quay or jetty alongside the vessel, supporting a belt conveyor on a horizontal telescope boom. The conveyor end can thus be positioned over almost any point of each hatch opening to load the coal.

Transportation in the terminal from unloader to stockpile and from stockpile to loading machines is by conventional belt conveyors usually laid out so as to be able to bypass the stockpile, thus allowing direct loading of coal as received from the terminals.

Self-unloading vessels, which reduce the need for investment in dockside equipment, have been operating on the Great Lakes in North America since 1950 and have recently come into use on the East and Gulf coasts of the United States. More extensive use may be possible in the future, but in order to justify the use of very large self-unloading vessels costing more than conventional ships, faster turn-around times at receiving terminals need to be achieved.

Usually, coal in bulk form is transported in ships as dry lumps with a top size of 32–50 mm and with a limited amount of fines. Total moisture content is usually well below 10 percent. The resulting bulk density of the coal is around 0.8 t/m³. Transporting the coal overseas in other forms would be more economic if by doing so more rapid or cheaper loading or discharge methods could be used or a higher energy density of the cargo achieved. Because the ocean transportation costs represent about 20–30 percent of the CIF (cost, insurance and freight) costs of the coal, there is strong incentive to search for ways to lower costs.

Among the alternatives under consideration are coal-water slurries and coal-oil slurries. Although the use of coarse slurries merely as a technique for rapid loading and discharging at the ports has been proposed, a more significant development would be the loading and movement of slurries transported by pipeline from inland mines to the ports. However, the use of water slurries for inland transportation may be restricted in some locations by the need for very large volumes of annual coal throughput (10 to 25 mtc/yr) to make them economic. Moreover, extensive development work remains to be completed before ocean movement of coal slurries can be considered economically viable.

The use of oil as the slurry medium may have some advantages in that the oil will burn as part of the fuel and thus will eliminate the need to remove excess water before combustion, as is required with coal-water slurries. Although slurries can be expected to play a role in coal transport by the year 2000, the bulk of the international coal trade is expected to be as dry cargo.

Ports and Terminals

Bulk terminals exist to facilitate the transfer of material between large-scale and small-scale methods of transport, and between different modes of transport. This is necessary to take advantage of the lower cost of large carriers for long distances while recognizing the constraints on the type and size of transportation that can be accepted by either producer or consumer.

Loading Terminals

A coal-loading terminal receives coal by train, barge, belt conveyor, or pipeline from producers inland. The coal is stored, blended, and intermittently dispatched in large sea-going vessels when a cargo of the required coal quality has been assembled. The coal is maintained in separate stockpiles for each coal specification required by customers. The loading systems for the coal vessels are designed for high capacity to decrease the loading time.

Receiving Terminals

A receiving terminal unloads coal from incoming ships and functions as a buffer between the large shipments arriving at irregular intervals and the more continuous delivery patterns required by the consumers. The facilities for unloading the sea-going vessels require a more sheltered berth than is necessary for the loading operation at the export terminal. At receiving terminals dedicated to one or more users whose installations are adjacent to the terminal, coal is stockpiled separately by specification or blended, when appropriate, before transfer to the users.

Transshipment Terminals

A transshipment terminal receives coal for user installations that are incapable of accepting direct delivery and that are often located at a significant distance from the unloading point. In this case the coal is shipped onward, and loading installations are provided for trains, trucks, barges, or coastal vessels. Such terminals may serve many customers and therefore, in addition to being more complex, usually have larger handling and storage areas than do simple receiving terminals.

Terminal Costs

The capital cost of terminals is determined largely by the size of ships to be handled. Not only the need for deeper water, requiring more expensive quay walls, more dredging, and longer trestles, but also the larger unloaders and higher unloading rates contribute to this cost.

In Table 6-5 typical costs are shown for investment "on shore" per metric ton of annual throughput capacity. Costs tend to be higher for receiving and transshipment terminals than for loading terminals. For receiving terminals that serve only one user the cost per metric ton capacity is generally higher because of the lower terminal throughput.

Table 6-5 Terminal Investment Cost
(1979 U.S. $/metric ton)

Type of Terminal	Throughput (million metric tons/year)	U.S. $ per metric ton annual throughput
Loading	20–25	6–8
Transshipment	20–25	8–10
Receiving	2–4	10–13

Because the costs for land and for dredging, quaywalls, trestles and jetties vary quite considerably with location, their construction and maintenance costs are not included in these figures. Terminal operating costs also depend much on local circumstances and therefore, only a broad indication can be given. The total annual cost of personnel, maintenance, power, and fuel ranges from 5 to 10 percent of capital investment.

Port Requirements for Future Coal Trade

To realize the projected increase in international coal trade, a significant expansion in terminal capacity will be required. The country reports contained in Volume 2 of this study outline many of the requirements and are summarized below.

Australia

Present export levels of around 40 mt/yr are handled through 6 ports only one of which, Hay Point, now has the capacity to handle 100,000 DWT vessels, though Newcastle will be able to handle 120,000 DWT ships in 1981. Increasing port capacity to the 170 mt/yr projected for the year 2000 would require both the upgrading and expansion of existing ports and the construction of several new ports. These developments would require corresponding investments in the rail systems linking the mines to the ports.

Canada

Exports of 14 mt/yr of metallurgical coal mined in western Canada are currently shipped from Vancouver. Existing terminal facilities are approaching capacity limits, but expansion of the deep-water Roberts Bank coal port is planned. Further growth in exports from the West Coast could be handled by the construction of a new coal port at Prince Rupert. Exports from eastern Canada could be increased by rebuilding the coal-loading facilities at Sydney, Nova Scotia, which currently handles small quantities of coal mined in Cape Breton. Canadian Pacific Coast ports could also be used to export coal from the northern Great Plains area of the United States if necessary.

Republic of South Africa

Coal is now exported mainly through the Richards Bay coal port, which is operating near its capacity of 24 mt/yr. Expansion to 44 mt/yr is now under consideration but a further growth in South Africa coal exports would require a new terminal. Richards Bay could also handle exports of coal mined in other countries of southern Africa, notably Botswana and Swaziland.

United States

The United States is currently the largest coal exporter—exports were 63 million tons in 1979. About 70 percent of current exports are metallurgical coal shipped primarily via the port of Hampton Roads, Virginia.

The basic conclusions of the U.S. port study are that (1) a significant expansion of U.S. port capacity would be required to meet coal export demand for all WOCOL cases; (2) the diversity and geographic distribution of steam coal exports would require new coal export facilities on the Gulf and West coasts, and also on the East Coast, where the bulk of the current exports originate; (3) capital requirements for the highest projected exports (350 mt/yr) could range up to $1 billion (U.S. 1979) for ports and $4 billion for inland transport; (4) regulatory and institutional considerations could be the most significant constraints to expanding existing ports or constructing new ones.

Western Europe

An established pattern of coal transportation exists around the coast of Europe and is capable of expansion to handle the projected coal movements in the WOCOL projections. Inland movement by river, canal, and rail can be expanded. There is, however, a lack of terminals capable of receiving coal in large bulk carriers, storing the coal, and transshipping it to its final destination. A number of sites are being studied as possibilities for development.

A large increase in coal imports into Scandinavia may be required, particularly if nuclear development proceeds at a reduced pace or if nuclear programs are severely curtailed. Several sites exist in both Denmark and Sweden that could be developed as receiving and transshipment terminals capable of handling 250,000 DWT vessels.

In northwest Europe the Amsterdam-Rotterdam-Antwerp (ARA) port system can be expanded to increase coal-handling capacity. Rotterdam is planning an expansion within the next few years in order to receive 250,000 DWT vessels.

On the west coast of the Federal Republic of Germany there are a number of ports capable of handling coal, the largest being Hamburg, which has the capacity to handle ships of 160,000 DWT and which also has substantial storage capacity.

In France, Le Havre and Dunkirk are the most important coal ports that could be expanded to receive large vessels. Considerable flexibility exists in expanding imports on the Atlantic coast, especially with new facilities in Nantes and Bordeaux. Facilities at Rouen, Cherbourg, Brest, Bayonne, and other ports could be developed to receive transshipment from Le Havre, Dunkirk, or the ARA ports.

There are several ports capable of significant expansion on the Mediterranean coast. Fos could be expanded from the present 3 mt/yr capacity to handle 10-20 mt/yr, and studies of such an expansion are being made.

The Italian coast offers several possibilities, and one site for a coal transshipment terminal has been selected. The Italian WOCOL report concludes that direct coal imports into five intermediate-sized ports receiving 100,000 DWT vessels may be suitable for handling projected future coal imports to the year 2000.

Japan and Other East Asian Countries

Capacity already exists in Japan to handle all foreseeable increases in imports of metallurgical coal. Imports of steam coal for power generation are likely to be initially handled in designated ports associated with specific power stations. At least two large "coal centers" are however also under consideration, and cost estimates indicate significant savings in total coal transportation cost.

Plans to handle the large volumes of imports projected for other East Asian countries are not well defined, but the availability of sites for terminals is not expected to prove a significant obstacle.

Details of these and other proposals may be found in Volume 2 of this study, but in general it is clear that the timing of port availability is likely to be a more critical factor in handling coal exports and imports than physical feasibility or cost of port expansions.

CHAPTER 7

COAL-USING TECHNOLOGIES

Established Technology for Direct Use — Coal Combustion — Coal Gasification — Coal Liquefaction — Lead Times — Coal Utilization Economics — Conclusions

Coal is an energy source that has evolved over several centuries from a heterogeneous mixture of vegetable origin. It consists mainly of carbon, hydrogen, and oxygen, with small quantities of nitrogen and sulfur. In addition, all coals contain inorganic ash-forming impurities. Because of its complex chemical composition, coal has a processing potential greater even than that of oil or natural gas. The products obtainable from coal range from electricity, gaseous and liquid fuels and chemical feedstocks, to metallurgical coke and activated carbon. But because coal is a solid and complex substance, a high technological skill is required to fully realize its potential.

Many developments in coal-using technology had emerged by the end of the last century. They comprised direct use of coal for heat and steam generation as well as conversion to other fuels. The first conversion process was retorting coal to char and coke, yielding gas and tar as by-products. The gas was used for lighting and heating; the tar, among other uses, formed the basis for the developing organic chemical industry. Early experiments were also carried out on complete gasification and liquefaction of coal. Gasification came into widespread industrial use in the 1920s. In Germany coal liquefaction gained vital importance during World War II. Liquefaction plants were also commissioned in other countries. However, the increasing availability of cheap oil and natural gas after World War II removed much of the incentive for any further work on liquefaction and gasification and brought about the eventual shutdown of coal liquefaction plants. Currently the only coal liquefaction plant of commercial size is located in the Republic of South Africa and has been operating since the mid-1950s.

Interest in coal as a fuel source intensified in the 1970s with the worldwide escalation of oil and gas prices and with increasing concerns about the future availability of oil and gas supplies. Because of its abundance and worldwide distribution, coal has again assumed increased importance as a versatile energy source that can meet a large share of the world's future energy needs.

Established Technology for Direct Use

In the industrialized countries the greater part of steam coal use is for the generation of electricity. Here the coal is dried, ground, and burned in a boiler to raise steam that drives a turbine to generate electricity.

Much progress has been made in improving the efficiency of electricity generation over the past few decades. In the 1950s the efficiency of power stations was commonly around 20 percent; now the best modern baseload stations achieve an efficiency of up to 40 percent, which has been made possible by increases in the temperature of the steam. However, this is now close to the maximum possible with conventional materials of construction.

Substantial developmental activity is in progress for further improvement of the control of particulate and gaseous emissions in order to meet increasingly strict environmental regulations. Considerable progress has been made in this area in recent years. With current technology, reductions of 99.8 percent in particulate emissions are achievable using either electrostatic precipitators or bag houses.

Most of the emphasis at present is on reduction of sulfur oxide emissions. Sulfur occurs in coal both in organic combination and in inorganic form, mainly as pyrite. The latter can usually be removed from the coal by physical cleaning techniques. The remainder is released as sulfur dioxide during coal combustion and requires removal, by law in many countries, from the flue gases. Wet scrubbers in which the sulfur is absorbed, usually by lime, are in commercial use, but these can create environmental problems of their own in the disposal of the resultant sludge. At an advanced stage of development are regenerative scrubbing systems that avoid the sludge problem and that result in the recovery of sulfur in its elemental and potentially useful form.

Some emphasis is being placed on the reduction of nitrogen oxide emissions, which arise from two sources: nitrogen contained in the coal, and molecular nitrogen in the air used to support combustion. Control of combustion conditions to lower peak temperatures and reduce excess air is effective to some degree in reducing nitrogen oxide formation. Furthermore, new processes are at an advanced stage of development for the chemical removal of nitrogen oxides from gaseous emissions.

Coal Combustion

In the residential and commercial sectors and in industry, most energy is used for heat production and is primarily supplied by oil and gas. In these sectors there is a large substitution potential for coal. To increase the incentive to use coal more extensively as a source of heat and steam in industry and for large-scale electric power generation, considerable effort is being directed to developing technologies that increase the efficiency and flexibility of coal combustion, extend the range of coals that may be used in a furnace, and help decrease environmental problems associated with coal combustion. The most important of these developing technologies are fluidized bed combustion and the production of stabilized coal-in-oil suspensions and their use in furnaces designed for oil, as well as magnetohydrodynamic (MHD) technology.

Fluidized Bed Combustion

Atmospheric-pressure fluidized bed combustion has been developed to the point that boiler vendors are prepared to extend plant guarantees for units of capacity up to 1 million pounds per hour of industrial steam generation. It is anticipated that this technology will be developed so that it can be utilized also for large central power-generating stations (in multiples of 200 to 300 MWe units) within the next 10 years. Pressurized fluid bed combustion units are at a substantially earlier stage of development.

Fluidization technology is well known and had its technical beginnings in the 1920s. It is currently in widespread industrial use in the petroleum-refining, chemical, and food industries (e.g., in catalytic crackers and in dryers for various materials). Fluidized bed systems are also used in waste incineration. These applications are characterized by moderate heat release rates. However, in coal combustion systems extremely high heat release rates are involved, requiring much more sophisticated equipment to control the bed temperature and to ensure full carbon combustion.

A fluidized bed combustor consists of a bed of crushed inert material, which may be a sulfur oxide acceptor such as crushed limestone, resting on a porous or perforated distribution plate. Air passing upward through this plate causes the bed to expand and in the correct velocity range confers on it the properties of a fluid. Crushed coal is fed into the fluidized bed, which has been initially preheated to operating temperature, where it readily burns at temperatures usually near 900°C (approximately 1,700°F). This temperature is considerably lower than the 1,300° to 1,500°C (2,400° to 2,700°F) usually experienced in conventional pulverized coal combustion. Water-carrying tubes are located within the bed, and these absorb the heat required for steam generation.

High heat-transfer efficiencies are achieved in the bed, despite the low temperature, that permit the absorption of sulfur oxides when crushed limestone is used in the bed. The coal ash, along with the spent sulfur oxide acceptor material, is continuously removed from the bed.

The fluidized bed combustion method offers a number of important advantages over conventional boilers. The absorption of sulfur oxides within the bed minimizes the problem of removing these substances from the flue gas. Furthermore, the low combustion temperature substantially reduces the formation of nitrogen oxides. It is also possible to use grades of coal containing very high quantities of ash—even coal preparation or colliery wastes can be burned. Fluidized bed units can also be smaller than conventional boilers of the same output, because the heat transfer to the water pipes running through the fluidized bed is more efficient than that in the conventional boiler.

Pressurized fluidized bed combustion units, currently under development, have the potential advantages of being even more compact and providing more efficient sulfur and nitrogen oxides reduction. Furthermore, the pressurized flue gas can be put through a combined cycle turbine for improvement of the efficiency of electric power generation. However, commercialization of this technology will likely require about 5 years longer than atmospheric fluidized bed combustion because many problems remain to be solved.

Coal-Oil Mixtures

In efforts to accelerate the increased use of coal there is interest—particularly in the United States and Japan—in the prospects for using pulverized coal dispersed in fuel oil. These coal-oil mixtures (COM) may use coal as a fuel in existing oil-fired installations to the extent of 20–50 percent. This is an old idea for which the feasibility has been reasonably well established in some short-term tests. The long-term stability of the mixture, the formation of slag from ash, the derating of boiler performance, and the installation of additional particulate flue gas-cleaning equipment and of coal-oil mixing facilities are factors that require careful consideration when planning to switch oil-burning equipment to COM firing. However, for new boiler capacity, the most efficient approach is to design for direct coal combustion.

Magnetohydrodynamic (MHD) Technology

MHD technology offers the potential of producing electricity directly from very hot combustion gases without using steam or gas turbines. This goal may be achieved by generating gas, in excess of 2,000°C, containing a seed material (e.g., potassium) that is readily ionized to render the gas electrically conductive (plasma), and passing this plasma with high velocity through a transverse magnetic field. The basic concept is very old, but many difficulties need to be overcome before this technology will become a practical reality. The problems include (1) the development of suitable high-temperature materials to contain the aggressive plasma, and (2) the efficient recovery of the seed material. If these problems can be solved, a combination of an MHD generator and a steam cycle can achieve efficiencies of about 45–55 percent compared with up to 40 percent for a conventional power station. Because of these difficulties, however, it is doubtful that this technology can be commercialized before the 1990s.

Coal Gasification

Gaseous fuels, because of the ease with which they can be handled as well as their wide range of applications as heat sources and chemical feedstocks, have an important role in industrial and domestic applications. In view of probable limits in the supply of natural gas and oil in the future, it is likely that coal gasification will come into increasing use to produce gas for use as a fuel and as a chemical feedstock.

Town gas production from coal traditionally has involved heating the coal in the absence of air—a process called pyrolysis or carbonization. Under these conditions gaseous and liquid products are produced from the volatile components in the coal, leaving a solid coke residue. The gas is used as a source of fuel and the liquid products as a feedstock for the chemical industry. Metallurgical coke for use in blast furnaces is also produced by carbonization, but the product of prime interest in that case is the coke rather than the gas.

Current interest in coal gasification aims at processes for total conversion of the coal to gas with little or no residue or by-products. A considerable amount of development on such coal gasification processes was carried out during the early decades of this century, and the processes now in general use originated during that period.

In principle all coal gasification processes involve the reaction of the coal with steam to form carbon monoxide and hydrogen. Because this reaction is highly endothermic, requiring a source of energy to make it go, some of the coal is reacted in parallel (simultaneously) with oxygen to provide the heat. The latter reaction produces carbon dioxide or, with adequate control of oxygen availability, more carbon monoxide. Thus the primary fuel gas derived from the coal consists principally of carbon monoxide and hydrogen in varying relative amounts depending on conditions.

Three grades of gas, in terms of their calorific or heating value, may be produced from coal. If air is used for the carbon (coal)-oxygen reaction, a low-calorific value gas is obtained (3.8–7.6 MJ/m^3) because the product gas is diluted by the nitrogen from the air used. It is not economical to send this grade of gas by pipeline over any distance, and thus it must be used at its source. It is of value for power generation and for industrial applications. Recent interest is motivated mainly by the possibility of producing an environmentally clean fuel from high-sulfur coal for use in power generation. Under coal gasification conditions the sulfur in the coal is primarily converted to hydrogen sulfide, which can be removed as elemental sulfur from fuel gas using well-established technology.

If oxygen, rather than air, is used for the carbon (coal)-oxygen heat-generating reaction, then a nitrogen-free, medium-calorific value gas is obtained ($10–16$ MJ/m^3). It is economical to send such gas by pipeline over reasonable distances and to use it directly for industrial and domestic purposes. This gas is of special value as feedstock for the chemical industry; as a source for hydrogen, methane, and methanol; and as a component in liquid hydrocarbon fuel production.

High-calorific value gas (greater than 21 MJ/m^3)—primarily methane—is produced by the so-called methanation reaction from carbon monoxide and hydrogen, i.e., from medium-calorific value gas obtained from coal as described above. When methane alone is the desired gas, i.e., substitute natural gas (SNG), it is necessary to first adjust the carbon monoxide-to-hydrogen molecular ratios to $1:3$ by reacting some of the carbon monoxide with steam to form hydrogen (and carbon dioxide, which is removed).

Interest in SNG production from coal is high because it offers the possibility of supplementing natural gas supplies in existing pipeline systems and for existing appliances. Another approach to SNG production from coal involves direct reaction with hydrogen to form methane. This process, known as hydrogasification, has not become a reality on a commercial scale mainly because of low conversion efficiency.

There are many coal gasification processes under development. The variety reflects the great heterogeneity and variability of coal as a raw material, the complex chemistry involved in conversion, the engineering problems associated with gasification, and the different qualities of product gas that are required.

Established Gasification Procedures

The operating parameters of established gasification processes are listed in Table 7-1. The airblown revolving-grate gas producer is the oldest. Formerly it was used widely, particularly in coking plants, where, under certain conditions, it delivered a fuel gas for operating the coke ovens so that the pyrolysis gas could be sold as town gas. This gasification method is again attracting attention for the production of low-calorific value gas for use as a clean fuel for coke ovens, boilers, and gas turbines. The process is comparatively simple and reliable. Tar and oil are obtained as by-products. Effluent water must be treated to remove phenolic contaminants.

Table 7-1   Operating Parameters of Some Commercially Available Coal Gasification Processes

	Revolving Grate Producer (airblown)	Lurgi Process	Winkler Process	Koppers-Totzek Process
Operating pressure in bar	1	25	1	1
Throughput in tons/hour	2–6	15	3–35	13–21
Feed coal	Mildly caking coal, coal lumps	Mildly caking coal, coal lumps	In particular lignite, crushed coal	All coals, pulverized coal
Gasifying agent	Air or oxygen/steam	Oxygen/steam	Oxygen/steam	Oxygen/steam
Carbon conversion (%)	98	99	90	90–96
Cold gas efficiency (%) (gasifier only)	72	75–85	75	70–77
Gas production m³ (STP)/hour	10,000–20,000	55,000	5,000–6,000	20,000–55,000
Specific oxygen consumption in m³ (STP)/ton coal	3,200 (air)	200–300	365	535
Specific steam consumption in tons/hour	Insignificant	1–1.4 (30 bar)	0.8 (low pressure steam)	0.24 (low pressure steam)
By-products	Tar, oil	Tar, oil	Coal fines, small amounts of tar	none

The Lurgi process (Table 7-1) is probably the best known. It was developed to the stage of large-scale commercial operation using subbituminous coals and lignites before World War II. In the early 1950s it was further developed to permit use of noncaking bituminous coals as well. It is now used in the Republic of South Africa to produce medium-calorific value gas for use as a fuel as well as an intermediate synthesis gas for the manufacture of motor fuels and various chemicals using Fischer-Tropsch synthesis. Several Lurgi plants are also in operation in Eastern European countries.

An advantage of the Lurgi process is that it operates at elevated pressure, making possible a substantial increase in throughput and saving compression costs for the raw gas sent to further processing. The Lurgi raw gas contains approximately 10 percent methane, which makes it particularly suitable for further upgrading to SNG. However, the Lurgi gasifier requires lump coal, and use of caking coals leads to a marked reduction in throughput. Tar and oil by-products make effluent water treatment necessary for removal of phenolic compounds. Further development of the Lurgi process is under way to increase throughput while reducing steam demands by raising the operating temperature, and to permit operation in a slagging mode for ash removal (British Gas Corporation and Lurgi joint development). Lurgi is also developing a further modification of its dry bottom gasifier to operate at higher operating pressure (i.e., 100 bar) jointly with Ruhrkohle AG (Federal Republic of Germany).

The Winkler fluidized bed gasification process is also in industrial use at present. However, because of temperature limitations it is restricted to highly reactive coals and is being run currently in several countries exclusively with lignite as feed. The reactor construction is very simple but the carbon carryover is high, and fly ash recirculation is required to achieve 90 percent carbon conversion. The process is extremely flexible in following load variations and can be built to achieve very large gasification rates. One of the first Winkler gasifiers was also successfully run with air rather than oxygen. All currently operating Winkler gasifiers are installed in fertilizer plants to produce synthesis gas for hydrogen production. A disadvantage of the gasifier in this application is the methane content of 1 to 5 percent in the raw gas, which requires an additional gas treatment operation not necessary in other methods of synthesis gas production. Recent development work on the Winkler process has involved raising operating temperature and pressure to higher values in order to improve gas composition and to permit the processing of coals with lower reactivity (low-volatility coals).

The Koppers-Totzek (KT) process can accept all kinds of coal from high-volatility types to low-volatility anthracite. The new four-headed gasifiers can be built with capacities up to 50,000 standard cubic meters per hour of raw gas. All ammonia plants built since 1960 that are based on coal use the KT process for the production of synthesis gas. Because of the very high operating temperature, the raw gas contains no methane. The process yields no tars or liquid hydrocarbon by-products, and thus no special effluent water treatment is required. However, the operation of the gasifier at atmospheric pressure usually requires product gas compression for further use, and the high operating temperature used to slag the ash requires a comparatively high oxygen intake. Carbon conversion, especially in less reactive coals, is not as good as in the Lurgi process, but operation of the gasifiers is quite simple and their dynamic response is particularly fast. Development work carried out by Shell and Koppers has led to the commissioning of a 150 tons/day pilot plant for pressurized gasification (the Shell Koppers process) aimed at removing the need to compress the product gas and at reducing the size and cost of gas-processing equipment.

A number of new processes are under development in several countries. Gasification at elevated pressure appears to be a common objective of all process developers because of the following advantages for this approach: increased throughput; reduced product gas compression requirements; smaller gas treatment equipment; physical gas washing processes applicable; and higher methane content in raw gas (at lower gasification temperatures only).

In some of the processes being developed in the United States, gasification is carried out in several steps in order to optimize chemical and thermal or engineering conditions for each step. These processes include the Hygas, Cogas, and Bi-gas methods, all of which have SNG as the objective. Another development route aims at relatively simple reactors for the production of synthesis gas at elevated pressure, for example, the Texaco and Saarberg-Otto processes as well as the Shell-Koppers method already mentioned.

Gasification requires that about 30 percent of the coal feed is burned with oxygen for the necessary heat supply. This coal is lost for conversion to gas. In countries with high coal production costs, for example in Europe, such losses cause the product gas cost to be high. In order to improve the economics of this heat supply operation, development is under way in the Federal Republic of Germany to supply heat for the process from high-temperature nuclear reactors. The potential of this technology is seen in a higher yield of gas (50 percent increase) per unit coal input and a substantially lower emission of pollutants (about 30 percent reduction) per gas unit produced. This combination also offers the opportunity to convert nuclear into chemical energy. Two pilot plants, one for gasification with steam and one with hydrogen, with a throughput of 4 tons/day each are operating successfully with the nuclear heat source being simulated.

The detailed engineering for a prototype plant with a high-temperature nuclear reactor of 750 MW thermal output has been started, and is currently expected to be available in the 1990s.

Gasification with air instead of oxygen is the main objective of the CO-Acceptor, U-Gas, Combustion Engineering, and Westinghouse processes. The purpose of these four processes is to produce low-calorific value fuel gas for use in combined cycle power plants based on coal. It is anticipated that this type of facility may come into widespread use as an environmentally acceptable means of providing fuel for the high-efficiency generation of electricity in some countries, such as the United States.

There are also studies of using synthesis gas produced by coal gasification as a feed to fuel cells for direct electric power generation, designed to achieve higher thermal efficiency for small-scale applications. However, no significant level of commercial operation is expected before the year 2000.

Underground Coal Gasification

Underground coal gasification (UCG) involves the in-place conversion of coal to gaseous fuels using air as the gasifying agent. This approach has a number of potential advantages. For instance, it eliminates the need for an underground mine labor force and minimizes the environmental problems associated with mining as well as with ash disposal and emissions to the atmosphere that arise in surface processing plants. On the other hand, UCG in its current state of development suffers from problems in process control, and the gas product requires upgrading for use. Because of its low heating value it also is uneconomical to transport over long distances and can serve only local markets.

The first experiments were carried out in the United Kingdom in the middle of the last century, and extensive work was done in the Soviet Union during the 1930s. However, since the 1950s very little further progress has been made because of lack of economic incentive and poor prospects for successful commercial application. In the early 1970s work was resumed in the United States and in Europe, aimed at gasifying brown coal as well as hard coal seams. It is not anticipated that this technology will have a significant impact on coal use before the year 2000.

Coal Liquefaction

For many uses, for example, motor fuels, there are no satisfactory alternatives to liquid fuels, and this has led to interest in technologies to convert coal to liquids. Moreover, the need to increase coal use in the generation of electric power while meeting environmental protection requirements has also encouraged interest in producing clean storable liquid fuels from coal.

In coals the atomic ratio of hydrogen to carbon is usually less than 1, whereas in petroleum-based fuel this ratio is in the range 1.5–2, depending on the grade. The conversion of coals to liquid fuels requires complex chemical modification with the introduction of additional hydrogen, and the parallel elimination of oxygen, sulfur, and nitrogen as well.

There are three general approaches to the production of liquid fuels from coal: (1) pyrolysis, (2) hydrogenation, and (3) gasification followed by conversion of the synthesis gas to liquids using Fischer-Tropsch or other technologies. In all approaches, hydrogen is required—either initially as in (2) and/or subsequently as in (1) and (2) for upgrading the crude primary coal-derived products. Thus, all coal liquefaction processes require some gasification of coal, or the carbonaceous residue from (1) or (2), with steam to produce the required hydrogen.

Pyrolysis

During pyrolysis, coal is destructively distilled at a high temperature in the absence of air. A redistribution of the hydrogen in the coal occurs to produce liquid and gaseous products as well as a significant hydrogen-depleted carbonaceous residue (char or coke). The Lurgi-Ruhrgas pyrolysis process has been commercially demonstrated. Current interest in this approach is mainly in the United States and Australia—the process at the most advanced stage of development being Toscoal and COED (char-oil-energy-development), from which has evolved the Cogas process. There is interest in establishing pyrolysis conditions to maximize liquid yields—these involve very rapid heating of pulverized coal (Federal Republic of Germany, the United Kingdom, the United States, and Australia). However, in all pyrolysis processes, the major product is a char that has potential direct use as a fuel and as a feedstock for fuel gas, synthesis gas, and further liquid fuel production. When high-sulfur coals are involved, the char product retains a high level of sulfur. Liquids derived from coal pyrolysis are complex tars that require hydrogenation to render them suitable for use as clean fuels.

Hydrogenation

Processes under development for the production of liquid fuels from coal by direct hydrogenation may be grouped into two categories—catalytic and noncatalytic. In the Federal Republic of Germany there has been a long history of experience in both approaches to coal hydrogenation. Bergius studied the principles involved before World War I. Commercial applications commenced in the 1920s, leading to the recognition in the United States that hydrogenation under pressure offered considerable potential for improving the yield and quality of motor fuels during the refining of crude oils. The first two commercial plants for this purpose were commissioned in the United States in 1930 and 1931 by Standard Oil of New Jersey (now Exxon). During World War II, 12 catalytic hydrogenation plants based on coal and tars using Bergius-Pier technology were operated in Germany. These provided 4 million tons annually, or one-third of German fuel oil requirements, mainly as automotive and aviation fuels.

These plants were shut down after the war because of the availability of cheap petroleum products. There are now no coal hydrogenation plants in commercial operation. However, a number of catalytic and noncatalytic processes are under active development in various countries.

The noncatalytic approaches are essentially directed to the high-efficiency conversion of coal to a product that is liquid at process temperatures with the minimum hydrogen consumption, hydrogen supply being a major cost in the operation of coal liquefaction processes. The objective is the removal of sulfur-containing carbonaceous and mineral residues by physical means to produce a fuel for direct use in utility and industrial boilers or as an intermediate for further upgrading to motor transport-grade fuels.

The most advanced of the processes in this category is the solvent refined coal process, which has been successfully piloted at a 50 tons/day scale in the United States. There are two versions: SRC-I, and SRC-II. The SRC-I mode produces a product that is a pitchlike solid at ambient temperatures but that is fluid at temperatures above 170°C. The instability of the product in the fluid state, coupled with its high viscosity, poses problems in the separation of solid residues. A tightening of legislation relating to sulfur emissions has led to the development of the SRC-II process with modifications in the process conditions, such as increased hydrogen pressure, to achieve a product that remains fluid at ambient temperatures with a lower sulfur content, at the penalty of increased hydrogen consumption. In 1979 engineering design and costing studies were under way in the United States for two 6,000 tons/day coal input plants—one to operate in the SRC-I mode, the other in the SRC-II mode. Japan and the Federal Republic of Germany are participating in the SRC-II project. With variations, the noncatalytic hydrogenation—that is, SRC—approach is also being actively pursued in Japan, the Republic of South Africa, and the United Kingdom. One of its by-products is especially useful in the metallurgical sector, in metallurgical coke and high-grade industrial carbon production.

In the direct catalytic hydrogenation approach to coal liquefaction the objective is to carry the conversion to the stage where the liquid product is suited for direct use as a heavy fuel oil or as a feedstock (syncrude) for refining to transport-grade liquid fuels. This requires the use of a catalyst and higher pressures than those involved in the noncatalytic hydrogenation approach.

In the original Bergius technology, extremely high pressures (up to 700 bar) and disposable iron catalysts were used. Recent developments have been aimed at decreasing the severity of conditions while maximizing conversion efficiency and minimizing hydrogen consumption. In the United States this has involved the use of expensive catalysts that must be recovered for reactivation and reuse, whereas in the Federal Republic of Germany interest in inexpensive "throwaway" catalysts remains.

In the United States the most advanced of a number of catalytic processes under development is the H-coal process, which copes with the catalyst management problem through special reactor design. This process has its origins in the H-oil process developed for upgrading high-sulfur heavy residual oils. A pilot plant having a coal feed rate of 600 tons/day when operating in the fuel oil mode, and 200 tons/day in the syncrude mode, has been constructed at Cattletsburg, West Virginia and is ready for commissioning. If the operation of this plant is successful, it is anticipated that a 20,000 tons/day (coal feed) commercial plant will be designed and constructed, but it could not be completed before 1985.

In the Federal Republic of Germany developments have reached a stage where a 200 tons/day coal feed plant is being constructed. The characteristics of the throwaway catalyst has permitted significant reductions in operating pressure. Detailed engineering design of commercial plants has commenced.

A second catalytic process attracting considerable interest is the Exxon donor solvent process, which seeks to avoid the complications of having the catalyst in direct physical contact with the coal, and of the separation of the liquid product from the solid residue. Here a hydrogen-donating solvent that has been recovered from the process is hydrogenated over a catalyst, then reacted with coal at elevated temperatures and under high hydrogen pressure. A pilot plant with 300 tons/day coal feed capacity is nearing completion.

Fischer-Tropsch Synthesis

Fischer-Tropsch technology using synthesis gas from coal was also used in Germany in World War II to produce liquid fuels. By the end of the war 9 Fischer-Tropsch plants were operating in Germany, 4 in Japan, 1 in France, and 1 in Manchuria.

The production of liquid fuels from coal by gasification to synthesis gas using Fischer-Tropsch technology is the only coal liquefaction process in commercial use today. In the Republic of South Africa a plant having an annual production of 240,000 tons/year of liquid fuels has been in operation since 1955. A second plant is now being commissioned that will increase production by 1.5 million tons/year. A third plant has been announced that will double the output capacity.

Coal gasification with Fischer-Tropsch technology has the advantage that it can be applied to low-grade high-ash coals. It has the disadvantage, when predominantly transport fuels are required, of poor selectivity. Progress is being made in improving the situation, and recent developments in the United States by Mobil Oil hold promise for further improvements. Mobil has developed, and to date tested at 4 bbl/day, a catalyst that allows quantitative conversion of methanol to gasoline. Because methanol can be selectively produced from synthesis gas using a suitable catalyst, there is renewed hope that a robust catalyst can be developed to permit the selective direct conversion of coal-based synthesis gas to gasoline. Several companies are currently testing such catalysts at a laboratory scale.

Lead Times

It is important to distinguish between the time needed to build a plant using commercially demonstrated technology and that needed for the development of new technology to the point at which it can be considered ready for commercial use. Even conventional power stations in the industrialized countries, which are now usually built in units of 400 to 750 MWe, require a period of 6 to 7 years for completion. About 4 years are needed for construction. In some cases, the total timespan from project initiation to commissioning of the plant can be up to 10 years, the extra time arising from the need for approvals from regulatory bodies and administrative procedures that often involve public hearings and long delays.

Experience shows that a period of about 30 years is necessary for the full-scale development of any major new processing technology from the initial concept. For new technologies still under development it is impossible to make precise estimates of future commercial-scale use. They are at different stages of development; they still face technical problems that must be solved. Furthermore, it is not known whether regulatory requirements in use in 1979 will still be valid when these processes are ready for commercial-scale plant construction.

For established gasification technologies, construction times are comparable with power stations. A similar situation exists for coal liquefaction. The well-established Fischer-Tropsch process used in the Republic of South Africa can be brought into production in about the same time as a gasification plant in other industrialized countries. The first full-size commercial plants using direct liquefaction technologies are not expected to come on line before 1990. Aggressive development programs currently planned for pilot plants and first commercial-scale facilities must be followed up by large commitments for commercial facilities if coal liquids are to make any significant contribution before 2000.

Coal Utilization Economics

A worldwide comparison of the costs of products from the different coal-using technologies is very difficult. Each country has its own particular cost and price structure. The financing requirements for the development of a project and the costing of plant operations (e.g., the depreciation guidelines or the different ways of handling the interest on and return on equity) have significant effects on the economic viability of a conversion project.

The most important cost factors in coal utilization processes are the capital costs of the conversion plant and the costs of the coal feedstock. Coal on an equivalent heating value basis is substantially less expensive than oil. Although coal is used in direct combustion for electric power generation, crude oil must be refined to produce a useable fuel. Consequently, refined fuel products such as Bunker C and distillate heating oils are typically used in oil-fired electric power generation. These fuels are somewhat more expensive than crude oil. Furthermore, as crude oil prices increase, residual fuels will tend to be upgraded in further refining to make more valuable products such as gasoline and jet fuel. During the next 10 years the availability of residual fuels derived from crude oil may therefore be expected to decrease.

Power Generation

A modern coal-fired power plant (700 MWe) without any flue gas desulfurization, with once-through water cooling and no other auxiliary plants, has an efficiency of about 40 percent. In the same modern power plant designed to meet stringent environmental standards (by the use of air cooling, complete flue gas desulfurization, and fine particle removal) the efficiency is reduced to about 34 percent. This results in an 18 percent increase in coal requirements (and therefore costs) for the production of the same amount of electricity. The capital requirements for a power plant are very dependent upon what equipment must be included to comply with environmental standards.

Gasification and Liquefaction

Only a limited number of processes for gasification and liquefaction are expected to be demonstrated and available for commercial use during the next 20 years. A comparison of the investment costs for a gasification or liquefaction plant for the same process in Australia, Canada, Japan, the Federal Republic of Germany, the Republic of South Africa, the United Kingdom, and the United States indicates that the capital requirements all fall within a spread of ±15 percent of an average value. Coal quality and labor, construction, material and equipment costs influence the total capital requirement. In addition, the operating costs depend on the specific financing arrangements. The cost of synthesis gas as a function of coal cost, shown in Figure 7-1, is based on the use of commercially demonstrated processes for plants ranging in size from 100 billion Btu/day to 250 billion Btu/day. These costs are based upon a 2:1 debt-to-equity ratio and on an 8 percent return on equity after a 50 percent tax. In many countries there is interest in SNG production for use in existing natural gas distribution systems. The methanation step adds from 50 cents

Figure 7-1 Cost of Synthesis Gas as a Function of Coal Cost

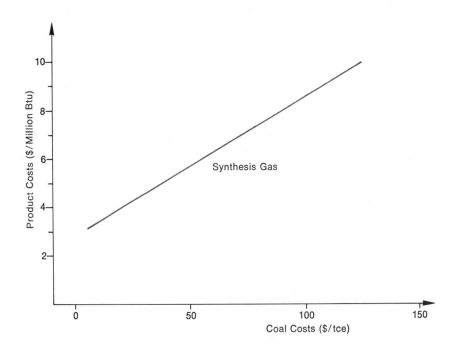

to $1 per million Btu to the gas cost depending upon coal cost and other factors.

The costs shown in Figure 7-2 are based on coal hydrogenation and provide a rough indication of the relationship between the potential costs of various liquid products and the dependence of these product costs on coal costs. The same economic bases were used as for Figure 7-1. Even though the cost of gasoline produced by Fischer-Tropsch technology is higher than that produced by hydrogenation, the total commercial value of Fischer-Tropsch technology includes additional values assigned to its by-products.

Figure 7-2   Costs of Liquefied Products as a Function of Coal Cost

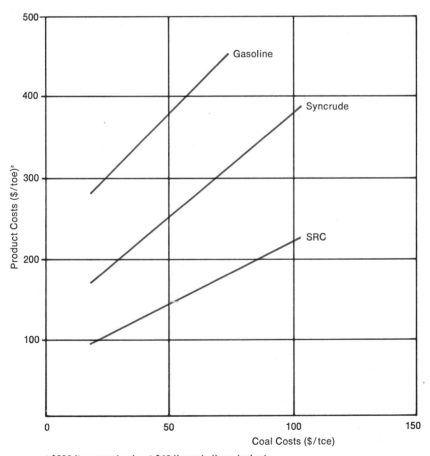

ᵃ $300/toe equals about $40/barrel oil equivalent.

Conclusions

Technology is available, and proved on a commercial scale, to permit increased use of suitable coals in supplying energy demands while complying with the requirements in most locations for protection of the environment. This is so not only for the combustion of coal as a direct source of heat and steam for the generation of electricity, and for industrial use, but also for the production of a wide range of gaseous and liquid fuels.

Economic, institutional, and other considerations—not technological ones—are delaying a more rapid buildup in coal use. Much effort is in hand to improve the efficiency, and hence economics, of established technologies for the direct use and conversion of coal and for meeting environmental protection requirements. Vigorous efforts are also under way for the development of new or modified processes for the conversion of coal to liquid and gaseous fuels.

The lead times involved in planning, building, and commissioning coal-based plants are significant, and decisions need to be made to build them to help meet future energy supply requirements at a time when petroleum-based fuels are still available at competitive prices. The initial impacts of greater coal use in reducing the demand for petroleum supplies will come mainly through replacement of oil and gas by coal for the generation of electricity and for the provision of heat and steam in industry.

CHAPTER 8

CAPITAL INVESTMENT IN COAL

Coal Chains — Capital Formation 1977–2000 Capital Estimates for Coal Supply — Capital Estimates for Coal-Fired Electric Power Stations — Factors Affecting Investment in Electric Power Systems — Potential Impediments to Financing Expanded Coal Supply — Conclusions

On the basis of the assumptions made in the WOCOL analysis, the production and use of coal is projected to grow at least 2.5–3 fold over the next 20 years, and world trade in steam coal is projected to grow 10–15 fold within the same period. This will require major capital investments in new and expanded coal mining, in coal-fired power stations and coal conversion facilities, in internal transport in both coal-producing and coal-using countries, in ports for the export and import of coal, and in ocean shipping.

Although the available data are limited, we have endeavored to estimate the order of magnitude of the aggregate capital expenditure required to achieve the projected rapid expansion in coal use.[1] It should be emphasized that these estimates are indicative and are included for illustrative purposes only. There are many difficulties in providing accurate estimates. For example, the requirement for capital investment in production, transportation, and distribution systems for both coal and electricity will be influenced by the existence of spare capacity in the relevant systems as well as by rates of retirement of the various kinds of facilities.

Coal Chains

We have found the concept of "coal chains" a useful device for visualizing the links in the system from mines to users, and for illus-

---

1. All estimates of capital requirements in this chapter are in 1978 U.S. dollars unless otherwise stated.

trating capital costs for each link, as well as the likely lead times for planning and construction. Figure 8-1 shows the generalized concept of domestic and international coal chains and their components. The shortest chain is to a mine-mouth power station in which the coal, after mining and preparation necessary to meet market requirements, is delivered directly to the power plant and the cost of transport is minimal. The longest chains involve ports and international ocean transport of up to 15,000 miles.

Figure 8-1   Generalized Domestic and International Coal Chains and Markets

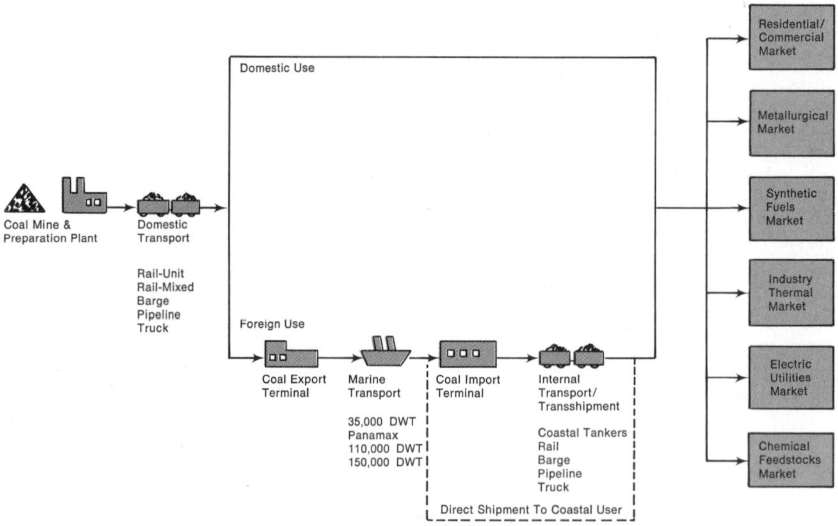

Several examples of coal chains have been produced by country teams and are included in their reports in Volume 2. Figure 8-2 shows a coal chain for delivery of U.S. western coal to electric power plants in the Far East, together with the implementation requirements at the various stages in the chain. The physical facilities, the capital costs, the labor required, and the lead times are shown for a coal flow of 5 mtce/yr. Such a coal chain from a mine in the United States to a user in Japan would involve federal leases for mining rights, private mines, private railroad companies, port facilities owned privately or by public authorities, privately owned ships, a receiving port in Japan that might be private or public, and private coastal shipping to an electric power plant that in most cases in Japan would be privately owned.

Figure 8-2 Illustrative Implementation Requirements for a Typical Coal Trade Chain—Western United States to Far East

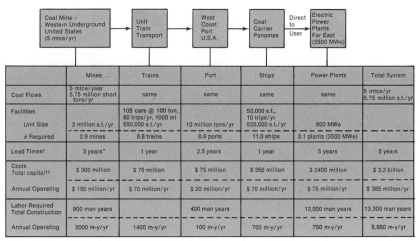

	Mines	Trains	Port	Ships	Power Plants	Total System
Coal Flows	5 mtce/year 5.75 million short tons/yr	same	same	same	same	5 mtce/yr 5.75 million s.t./yr
Facilities		105 cars @ 100 ton, 60 trips/yr, 1000 mi		52,000 s.t., 10 trips/yr		
Unit Size	2 million s.t./yr	650,000 s.t./yr	10 million tons/yr	520,000 s.t./yr	800 MWe	
# Required	2.9 mines	8.8 trains	0.6 ports	11.0 ships	3.1 plants (2500 MWe)	
Lead Times†	3 years*	1 year	2.5 years	1 year	5 years	5 years
Costs Total capital††	$ 300 million	$ 75 million	$ 75 million	$ 350 million	$ 2400 million	$ 3.2 billion
Annual Operating	$ 150 million/yr	$ 70 million/yr	$ 20 million/yr	$ 70 million/yr	$ 75 million/yr	$ 385 million/yr
Labor Required Total Construction	900 man years		400 man years		12,000 man years	13,300 man years
Annual Operating	3000 m-y/yr	1400 m-y/yr	100 m-y/yr	700 m-y/yr	750 m-y/yr	5,950 m-y/yr

* Lead times for coal mines assume that resource exploration has previously been completed and refer only to bringing the first phase of capacity into production.
† Lead times for actual project execution after all permits granted.
†† January 1978 dollars, includes escalation and interest during construction.

In the case shown in Figure 8-2 the capital cost of developing the mines, internal transport, and export port facilities amounts to $450 million, and the necessary ocean ships cost another $350 million. However, the largest item of capital cost in this chain is the $2,400 million cost of the power plants in the Far East, assuming that new facilities are required and that the power plants are operating at a 65 percent capacity factor. Thus in this case the capital cost of a new user facility is 75 percent of the total cost of the coal chain; and the 5-year lead time required for its construction is the longest in the chain and therefore determines the lead time of the total system.

Other typical coal chains illustrated in Figures 8-3 to 8-6 show similar dependence of the whole chain on early decisions concerning the power plant and the dominance of its capital cost in the total cost. For deep mines in Europe, however, the lead time required to develop a new mine may be even longer than for the power station.

The illustrative coal chain diagrams can also be used to show a cumulative build up of the cost per ton at each stage from mine to

Figure 8-3(a) Illustrative Implementation Requirements for a Typical Coal Chain*—Australia to Far East, Recent Project

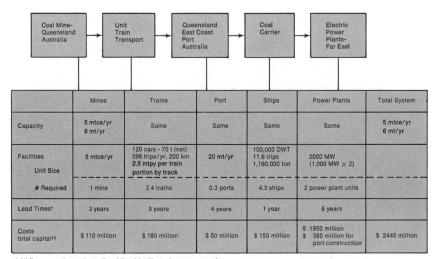

	Mines	Trains	Port	Ships	Power Plants	Total System
Capacity	5 mtce/yr 6 mt/yr	Same	Same	Same	Same	5 mtce/yr 6 mt/yr
Facilities Unit Size	5 mtce/yr	120 cars - 70 t (net) 298 trips/yr, 200 km 2.5 mtpy per train portion by track	20 mt/yr	100,000 DWT 11.6 trips 1,160,000 ton	2000 MW (1,000 MW × 2)	
# Required	1 mine	2.4 trains	0.3 ports	4.3 ships	2 power plant units	
Lead Times†	3 years	3 years	4 years	1 year	5 years	
Costs total capital††	$ 110 million	$ 180 million	$ 50 million	$ 150 million	$ 1950 million $ 385 million for port construction	$ 2440 million

* All figures are "guesstimates" and listed for illustrative purposes only.
† Lead times for actual project execution after all permits granted.
†† January 1978 dollars, includes interest during construction.

Figure 8-3(b) Illustrative Implementation Requirements for a Typical Coal Chain—Australia to Far East, Potential Project

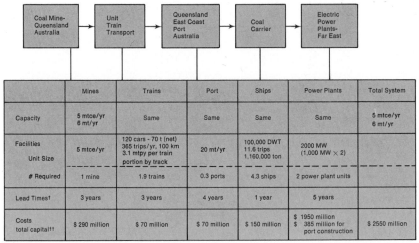

	Mines	Trains	Port	Ships	Power Plants	Total System
Capacity	5 mtce/yr 6 mt/yr	Same	Same	Same	Same	5 mtce/yr 6 mt/yr
Facilities Unit Size	5 mtce/yr	120 cars - 70 t (net) 365 trips/yr, 100 km 3.1 mtpy per train portion by track	20 mt/yr	100,000 DWT 11.6 trips 1,160,000 ton	2000 MW (1,000 MW × 2)	
# Required	1 mine	1.9 trains	0.3 ports	4.3 ships	2 power plant units	
Lead Times†	3 years	3 years	4 years	1 year	5 years	
Costs total capital††	$ 290 million	$ 70 million	$ 70 million	$ 150 million	$ 1950 million $ 385 million for port construction	$ 2550 million

† Lead times for actual project execution after all permits granted.
†† January 1978 dollars, includes interest during construction and including necessary infrastructure.

206

Figure 8-4 Illustrative Implementation Requirements for a Coal Chain—Australia to Netherlands*

	Coal Mine NS Wales Australia	Unit Train Transport	NS Wales Port Australia	Coal Carrier	Rotterdam Port Netherlands	Barge Transport Nijmegen	Power Plants Netherlands	Total System
	Mines	Trains	Ports	Ships	Port	Barges	Power Plants	Total System
Coal Flows	5 mtce/yr 5.6 mt	5.6 mt	5.6 mt	5.6 mt	5.6 mt	5.6 mt	5.6 mt	5.6 mt
Facilities	7.5 mt raw coal 3.75 mt saleable 1.5 mines	80 km 42 cars 72 t (net)/car 3.8 trains	30 mt/year 0.2 ports	110,000 DWT 13 ships	25 mt/yr 0.25 terminal	2.8 pusher tugs 28 barges	600 MWe 4.6 plants (2700 MWe)	
Lead Times †	3 years	3 years	2.5 years	¾ years each	2.5 years	¾ years each pusher tug	5 years each	5 years
Costs Total Capital ††	$310 million	$50 million	$60 million	$350 million	$40 million	$34 million	$2300 million	$3144 million

* All figures are estimates and listed for illustrative purposes only.
† Lead times for actual project execution after all permits granted.
†† 1979 US$, includes escalation and interest during construction.

207

Figure 8-5 Illustrative Implementation Requirements for a Typical Coal Chain—Canada to the Netherlands

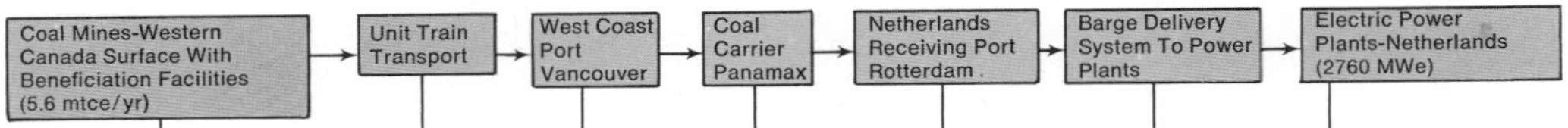

Coal Mines-Western Canada Surface With Beneficiation Facilities (5.6 mtce/yr) → Unit Train Transport → West Coast Port Vancouver → Coal Carrier Panamax → Netherlands Receiving Port Rotterdam → Barge Delivery System To Power Plants → Electric Power Plants-Netherlands (2760 MWe)

	Mines	Trains	Shipping Port	Ships	Receiving Port	Barge Delivery	Power Plants	Total System	Total Canada
Coal Flows	5.6 mtce/yr. 7.0 million tons/yr.	same	same	same	same	same	same	5.6 mtce/yr. 7.0 million tons/yr.	5.6 mtce/yr. 7.0 million tons/yr.
Facilities Unit Size	3.5 million tons/yr.	100 cars of 90 tons 50 trips/yr. 1200 km 450,000 tons /yr. †60km extension at $1.2 million/km	Expansion of existing facility at $10/annual ton 20 million tons/yr.	75,000 tons 7 trips/yr. 525,000 tons/yr.	25 million tons/yr.	-Loading facilities at receiving port -Unloading facilities at power plant 2.8 pusher tugs 28 barges	600 MWe 4.6 plants MWe	2760 MWe	2760 MWe
# Required	2 mines	16 trains	0.4 ports	13.3 ships	0.3 ports				
Lead Times †	3 years*	1.5 years	2 years (expansion only)	2 years	2.5 years	1.5 years	5 years	5 years	3 years
Costs Total Capital ††	$525 million	$216 million	$70 million	$466 million	$100 million	$34 million	$2300 million	$3.7 billion	$0.8 billion
Annual Operating	$154 million/yr.	$98 million/yr.	$18 million/yr.	$98 million/yr.	$21 million/yr.	$27.5 million/yr.	$400 million/yr.	$817 million/yr.	$270 million/yr.
Labor Required Total Construction	500 man years	150 man years	300 man years		500 man years		6900 man years	8350 man years	950 man years
Annual Operating	1000 m-y/yr.	1200 m-y/yr.	25 m-y/yr.	1050 m-y/yr.	75 m-y/yr.	50 m-y/yr.	700 m-y/yr.	4100 m-y/yr.	2225 m-y/yr.

† Lead times for actual project execution after all permits granted.
†† January 1978 dollars, includes escalation and interest during construction.
* Lead times for coal mines assume that resource exploration has previously been completed and refer only to bringing the first phase of capacity into production.

208

Figure 8-6 Illustrative Implementation Requirements for a Typical Coal Chain—Western Canada to Denmark[2]

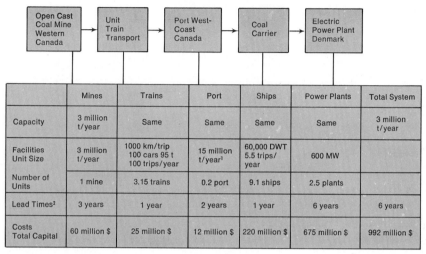

	Mines	Trains	Port	Ships	Power Plants	Total System
Capacity	3 million t/year	Same	Same	Same	Same	3 million t/year
Facilities Unit Size	3 million t/year	1000 km/trip 100 cars 95 t 100 trips/year	15 million t/year[1]	60,000 DWT 5.5 trips/ year	600 MW	
Number of Units	1 mine	3.15 trains	0.2 port	9.1 ships	2.5 plants	
Lead Times[2]	3 years	1 year	2 years	1 year	6 years	6 years
Costs Total Capital	60 million $	25 million $	12 million $	220 million $	675 million $	992 million $

[1] Extension of existing port.
[2] Lead times for actual project execution after all permits granted.

Figure 8-7 Unit Costs in an Illustrative Coal Trade Chain— Open Cast Mine in Western Canada to Electric Power Plant, Denmark

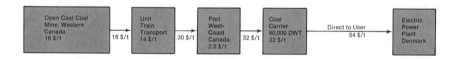

Figure 8-8 Unit Costs in an Illustrative Coal Trade Chain— Underground Mine, South Africa to Electric Power Plant, Denmark

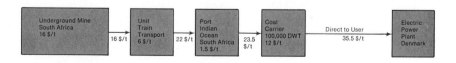

(2) Both Figures 8-5 and 8-6 show chains from a surface mine in western Canada to Western Europe. Figure 8-5 is for 5.6 mtce/yr; Figure 8-6 is for 3 mtce/yr. The substantial differences in cost estimates reflect different circumstances and views of different WOCOL teams.

209

user (Figures 8-7 and 8-8). In these international coal chains, mine production cost is seen to be less than half of the CIF cost, inland and maritime transport accounting for the remainder. In domestic coal chains, of course, transport cost may be a much lower share of final costs although there are exceptions. In the western United States, for example, where relatively low-calorific value coal is transported over long distances to users, transport costs may amount to 3 times the mine-mouth cost of the coal.

Chapter 6 presents more detailed analysis of coal trade routes, distances, port costs, and other aspects of ports and maritime transport. A range of unit capital costs for ports is shown in Table 6-5 (per ton of coal equivalent of annual throughput). Capital costs for loading terminals are $6–8/tce; for a transshipment terminal $8–10/tce; and for a small-scale receiving terminal $10–13/tce.

In the coal chains, illustrated in Figs. 8-2 to 8-6, estimated ship costs range from $25 to $35 million per ship for ships ranging in size from 60,000 to 110,000 DWT. The figure from Chapter 6 is $45 million per 100,000 DWT ship. Port costs range from $4 to $13 per annual ton. Such ranges illustrate the range of estimates of ship costs and the variety among coal chains. The two chains of western Canadian coal to Europe (Figures 8-5 and 8-6) assume Panamax-size vessels (60-75,000 DWT) for passage through the Panama Canal. The low port cost in Figure 8-6 assumes expansion of an existing port. Port costs in the other chains are $8–13 per annual tce of capacity and lie in the range of the figures in Chapter 6.

Ownership of Coal Facilities in the Coal Chain

Ownership of links in the coal chain will affect the way expansion is financed. Unlike oil supply chains prior to 1970, ownership of individual components of coal chains is largely fragmented, and the financial capabilities at each link vary widely.

Coal mines and preparation plants are generally publicly owned in some countries including India, the People's Republic of China, France, Poland, and the United Kingdom, and privately owned in others such as Australia, Canada, the Republic of South Africa, and the United States. Ownership of the connecting links—railways, pipelines, ports, barges and ships—is mixed. In most countries railways are publicly owned, important exceptions being the United States system and about half of the Canadian system. Ports are mostly owned by public entities, the principal exceptions being in the United States and the Republic of South Africa. The world's shipping fleet is largely privately owned except in centrally planned economies. In most countries electric power stations are wholly or partly publicly owned, exceptions being the Federal Republic of Germany, Japan, and the United States. Even where private ownership is widespread, as in the United States or Japan, parts of the system are also publicly owned.

Power stations have historically been largely financed from domestic sources, principally through direct government supply of capital or sale of bonds of public entities, and usually in the domestic market. In most countries this pattern is expected to continue, except where the size and speed of build up of coal-fired power stations (and possibly of synfuels plants) may exceed the capacity of domestic capital sources and require financing in the international capital markets.

Although the largest part of the capital investment in a coal chain is in the user facility—for example, a power plant—the capital cost of the remaining links is still substantial for some coal-exporting countries. However, the WOCOL country reports do not indicate that insurmountable difficulties are expected in financing coal supply chains as long as there are assurances of adequate long-term demand from buyers.

Capital Formation 1977–2000

The summary worksheets from each WOCOL country team (Appendix 1) include 1977 gross national product (GNP) in 1978 U.S. dollars and the rates of economic growth projected for 1977–2000. Columns 1 and 2 of Table 8-1 tabulate this data for the WOCOL countries in the Organization for Economic Cooperation and Development (OECD). From such data an estimate of the aggregate GNP for the years 1977–2000 can be computed, as shown in column 3.

The International Monetary Fund publishes historical data on gross capital formation for individual countries.[3] Average values for the years 1973–1977 for the WOCOL countries in OECD are given in column 4. Projecting the same averages forward for the years

---

3. *International Financial Statistics* 32(11) (November 1979).

1977–2000, the aggregate capital formation for the years 1977–2000 is calculated in column 5.

Table 8-1 Estimated Capital Formation For WOCOL Countries in OECD 1977–2000

Country	(1) 1977 GNP[a] ($ Billions)	(2) 1977–2000 GNP Growth Rate[a] (%/yr.)	(3) 1977–2000 Aggregate GNP ($ Billions)	(4) Average Gross Capital Formation % GNP[b] 1973–1977 (%)	(5) 1977–2000 Aggregate Capital Formation ($ Billions)
Australia	$ 95.6	3.5%/yr.	$ 3,370	23.53%	$ 792
Canada	197	3.5	6,950	23.09	1,604
Denmark	52.1	2.4	1,675	23.75	398
Finland	30	3.0	990	29.04	287
France	381	4.0	14,900	23.46	3,500
Federal Republic of Germany	514.1	2.8	16,590	21.74	3,606
Italy	223.8	3.3	7,685	20.77	1,596
Japan	912	4.6	37,250	32.90	12,255
Netherlands	120	3.0	4,055	21.16	858
Sweden	78	2.6	2,485	21.27	529
United Kingdom	257	3.0	8,510	19.33	1,644
United States	1,900	2.8	63,050	17.25	10,876
Total[c]	4,760	3.4	171,000	23	38,000

[a] From Appendix 1, WOCOL Case B.
[b] Source: IMF.
[c] Totals are rounded.

In the following sections of this chapter this figure of gross capital formation for these countries has been used as a rough guide to the magnitude of capital that will be available in capital markets. There are of course many demands for capital, and some countries have traditionally obtained part of their capital in international capital markets. Such patterns may be expected to continue in the future, although the extent to which the various markets are likely to be available or appropriate will fluctuate over time.

Capital Estimates for Coal Supply

The illustrative coal chains shown in Figs. 8-2 to 8-6 provide a basis for making rough estimates of the capital costs of coal supply systems from mine to user. In Table 8-2 the capital cost of components of the six illustrative coal chains are shown, converted to a common basis of capital cost per annual ton of new capacity ($/tce/yr). Actual investments in the coal supply chains will vary greatly among countries.

The principal capital need for expanding production will be in coal-producing countries and for ships. For some coal importers, existing ports may be adequate to handle projected coal tonnages with some expansion. Other importers may need to build new ports to handle large, deep-draft ships and to provide for transshipment to users. In either case the capital investments for coal receivers are relatively small in the context of total capital requirements for coal chains—of the order of $12 per annual tce of port capacity.

On the other hand, producer-exporters face substantial investments in mines, inland transport, and export ports. In addition a substantial investment must be made in ships. Without backhaul cargoes and assuming ten round trips per year, and if all ships were new, about 1,000 ships of 100,000 DWT would be needed for international transport of 1 billion tons per year. At a cost of $35–45 million each, the total cost for ships over the 20-year period would be $35–45 billion ($1979 U.S.).

Table 8-2 Capital Costs in Coal Supply Chains ($1978 US per annual tce)

Figure No.	Source	Destination	Mines	Inland Transport	Ports— Loading and Receiving	Ships	Total
8-3a	Queensland, Australia (recent project)	Far East	22	36	20	30	108
8-3b	Queensland, Australia (potential project)	Far East	58	14	28	30	130
8-4	New South Wales, Australia	Netherlands	62	17	20	70	169
8-5	Western Canada	Netherlands	94	45	30	83	252
8-6	Western Canada	Denmark	20	8	8	73	109
8-2	Western United States	Far East	60	15	30	70	175
	Average		53	23	23	59	158
	Range		20–94	8–45	8–30	30–83	108–252

Source: Figures 8-2 to 8-6. Excludes user facilities.

In endeavoring to estimate capital costs, it needs to be borne in mind that it is physical tons that must be mined, transported, and shipped, and physical tons will be somewhat greater than tce depending upon the calorific value of the coal.

From Table 8-2 the average capital costs per annual tce of new capacity for these several chains are as follows in U.S. 1978 dollars.

- Mines $53.   per annual tce
- Inland Transport $23.   per annual tce
- Ports $23.   per annual tce
- Ships $59.   per annual tce

Coal use in the WOCOL countries in OECD is projected to increase to the year 2000 by about 2,000 mtce/yr. Assuming that these average illustrative costs are broadly representative for other producers and exporters, this would require an investment for mines and internal transport of $76 per annual tce or a total of $152 billion for an increase of 2,000 mtce. The additional investment for export and import ports and ships to handle an increase in world coal trade of 600 mtce would be $82 per annual tce or a total of $49 billion. Although such sums are large in absolute terms they are a very small percentage—less than 1 percent—of the estimated aggregate capital formation of about $38,000 billion for these WOCOL countries during the period to the year 2000 (see Table 8-1).

The projections for the centrally planned economies indicate a coal increase of about 1,500–2,000 mtce by the year 2000—about the same as for the OECD countries—a large part of which is in the People's Republic of China and the Soviet Union. For the less developed countries (LDCs) the increase may be about 500–750 mtce/yr or more. If capital costs in those regions are comparable with costs in the OECD, total capital requirements could be somewhat greater than for the OECD countries.

Capital Estimates for Coal-Fired Electric Power Stations

From the coal chains illustrated in Figures 8-2 to 8-6 it is clear that about 75 percent of the cost of a complete coal chain (including the user) is for the user facilities. Over 90 percent of the capital cost of coal for coal importers may be for the using facility.

Appendix 1 includes the WOCOL country team estimates of electricity growth rates and coal-fired electric capacity projected to be built by the year 2000. Table 8-3 is a tabulation of such estimates for the countries in the OECD. Using a figure of $1 billion $1978 U.S./ GWe, the total capital cost of new user facilities totaling about 740 GWe would be $740 billion, excluding power transmission and distribution. This is a very substantial sum. In fact in many countries electric power systems have usually constituted the largest item of capital expenditure. Nonetheless, the financing of such facilities appears to be within the potential combined capacities of domestic and international capital markets over the period covered by the study.

Table 8-3 Net Additions of Coal-Fired Power Plants in OECD Countries 1977–2000

Country	Electricity Growth Rate % / yr[a]	Net Additions to Coal-Fired Electric Capacity GWe[a]
Australia	5.7	48.6
Canada	4.1	49
Denmark	4.4	10.0
Finland	3.3	3.6
France	5.2	19.7
Federal Republic of Germany	3.5	27.0
Italy	3.6	20.6
Japan	4.5	48.0
Netherlands	2.9	15.8
Sweden	1.7	11.6
United Kingdom	2.5	10.0
United States	4.1	423.
Other Western Europe	4.7	53.0
OECD Total	3.9	740

[a] From Appendix 1 Case B.

In assessing the size and relative importance of investment in new coal-fired power stations, the investment required for coal should be compared with that required if other available fuels were to supply the same end use. If electricity demand grows in the manner shown in WOCOL country studies, a substantial program of building new power stations will be required. Financing must be found for the full cost of the power station, whatever the fuel, and for any related transmission and distribution facilities that may be required to handle system expansion. Thus, the relevant cost to attribute to the use of coal is the marginal difference between the coal-fired power station and the oil-fired or nuclear alternative.

There are wide variations in the constructed cost of coal-fired power plants in various localities in the world arising not only from differences in unit sizes, site acquisition costs and construction conditions, local labor costs, equipment costs and the cost of its delivery, different import and tax costs, etc., but also from varying environmental protection requirements and the scope of facilities included in the plant cost. Further, there are many different ways of accounting for the various owner costs including interest costs during construction as well as the impact of escalation on these costs.

In this example we have sought to use costs at the level experienced in 1979 without adjustment for past or anticipated escalation —the cost of a facility as though it were possible to build it all at once. To this have been added the various owner costs (administration, engineering, site services, etc.) and interest at rates reflecting relatively low rates of inflation rather than those prevalent in every part of the world at present.

Even though on such a basis a large, modern coal-fired power plant utilizing stack gas scrubbers and cooling towers might cost from $600 to $750 per kw in the United States it is possible to have costs perhaps twice as high for unusual requirements (for example, smaller plants, dry cooling, NO_x removal by chemical means). We have used a broad average of $1,000 per kw in 1978 U.S. dollars for the purpose of estimating financial requirements for additional coal-fired power plants under a wide variety of conditions.

Factors Affecting Investment in Electric Power Systems

Delays in ordering new coal-fired power plants may delay the whole coal supply system, because activation and financing of the coal production and delivery systems depends upon confidence that the coal can be sold at a price that justifies the investment, and in some cases upon the development of long-term contracts.

Athough in many countries, electric power systems have long been the largest single item of total capital investment, they may take a larger proportion in the future as a result of increases in power plant costs. Each country therefore will need to consider how the scale of capital needs for its electric systems affects its ability to raise capital from domestic and international capital and credit markets.

A reduced growth rate for electricity since 1974, which seems likely to continue, leaves some electric systems with excess reserve capacity. These systems may have difficulty in obtaining capital to substitute coal-fired plants for existing oil plants despite a lower electricity cost from coal plants. This problem is complicated further by new oil-fired plants, some of which are still under construction. Moreover in some countries electric systems have made large investments, and are incurring heavy financing charges, for nuclear plants that are not yet completed or, in some cases, although completed are not permitted to operate. Where this has occurred, there may also be a restricted ability or willingness to raise the capital needed to build additional coal-fired plants. This implies that in some countries, the United States for example, a national policy to encourage substitution of coal for oil may need to be accompanied by measures that take account of the financial constraints on electric systems so that the necessary financing can be obtained.

Potential Impediments to Financing Expanded Coal Supply

Coal consumers, especially those utilizing coal for power plants, will want to identify and obtain long-term access to adequate supplies of coal before committing to substantial investments in new capacity. Coal producers, on the other hand, in common with those responsible for coal transportation systems, will be anxious to ensure that they are not committed to substantial investment programs in advance of the reasonably assured development of adequate long-term demand for coal. Although to some extent this apparent dilemma of circular uncertainties is avoided by the present existence of some excess coal supply capacity, particularly in the United States (which accounts for the largest share of the total investment contemplated in the WOCOL countries), it may be useful in this context to briefly identify potential impediments to financing the expansion of coal supply on the scale required by the WOCOL projections.

The extent to which potential impediments exist will, of course, vary from one country to another and from one project to another, assuming that the usual technical, geological, and managerial criteria can be met. However, a reasonably representative list of the more significant problems for both lenders and investors would include the following items: (1) the need to develop a reasonably assured demand for coal; (2) uncertain political environments that may increase the operating and capital costs for a project (e.g. through changes in environmental control legislation) or that might disrupt the production from the mine or affect security of tenure; (3) insufficiently attractive investment environments (governmental incentives, assurances of long-term tenure, etc.) to attract the levels of investment required to promote the rapid development of coal-mining and transportation capacity; (4) the unavailability or limited availability of reliable long-term purchase contracts with credit-worthy buyers, including pricing mechanisms offering a reasonable expectation of generating sufficient revenue to cover debt service and provide a reasonable return to investors; (5) excessive capital costs caused, for example, by the inadequacy of existing infrastructure required to implement a project, which may impair its economic and financial viability; (6) inadequate capacity of some domestic capital markets in circumstances where the credit of the project's sponsors, including national governments, is insufficient to tap into international markets and where long-term sales contracts cannot be arranged to the extent necessary to improve the credit standing of the project; and (7) nonavailability, in some LDCs, of an assured supply of foreign exchange required to service the potential project debt.

Turning to the question of the size of the total financial requirement projected by the WOCOL study, the scale of funding contemplated does not appear unrealistically large when compared with the potential combined capacity of the world's domestic and international capital and credit markets.

Even though a detailed analysis of the historical development and capacity of these markets in the context of an assessment of their future ability to accommodate the levels of expenditure projected in the WOCOL study is not intended, it is of interest to note the recent growth of the international capital markets shown in Table 8-4. In particular, it is of interest to note the steady growth of LDC borrowings as a percentage of total borrowings. To the extent that recent rapid rates of growth in the international capital markets are not sustained, or that these markets come under increasing pressure from competing demands unrelated to coal, greater reliance will need to be placed on domestic capital markets. This may not of itself pose any particular difficulty for most sponsors of coal projects in developed countries. However, where funds to be raised are large in comparison with a borrower's or guarantor's marginal debt capacity as perceived by the financial community, some individual producers and consumers, even in developed countries, may in the normal course be expected to encounter significant impediments to their efforts to fund new projects, and some investors will require additional incentives and assurances before committing a substantially increased portion of their resources for coal-related projects.

Table 8-4 Funds Raised by Borrowers in the International Capital Markets[1] 1975–1978

	International Bonds		Publicized Eurocurrency Credits*		Total	
	All Categories $ millions	LDCs %	All Categories $ millions	LDCs %	All Categories $ millions	LDCs %
1975	10,520	4	20,553	61	31,073	42
1976	15,368	8	28,703	60	44,071	42
1977	19,484	14	34,185	59	53,669	43
1978	15,872	20	72,025	53	87,531	47

[1] Source: World Bank Publication EC-181/791, *Borrowing in International Capital Markets*, July 1979.
* The estimates in Table 8-4 understate the true volume of these markets to the extent that they do not include unpublicized credits and issues.

Some problems may also be encountered in the case of the LDCs, which appear likely to continue to run substantial balance of payments deficits. In order to fund capital investment programs on the scale contemplated, a number of LDCs will need to raise substantial amounts of foreign debt to avoid undue pressure on their domestic capital markets; others have virtually no internal capacity to finance significant expansion of coal production and use and must look to the World Bank and other outside capital. The relatively small size of the international capital markets (see Table 8-4) may result in the danger that unless the developed countries utilize their domestic markets to the fullest extent practicable, LDC borrowers may be crowded out of the international markets and become increasingly reliant upon aid and the international multilateral development banks.

Conclusions

These estimates of the order of magnitude of the capital investments needed to achieve the level of coal expansion projected which would involve tripling coal use and expanding world steam coal trade 10–15-fold, lead to the following conclusions.

1. Three-quarters or more of the total capital investment lies in the user facilities—principally electric power plants. Lead times for planning and building such facilities are generally longer than any other part of the coal supply chain. Electric power plant investments will be large, whether coal or another fuel is used, if growing electricity demand is to be met.

2. Of the expected coal use in the OECD, about 70 percent is for domestic consumption involving mines and inland transport. The remainder—about 30 percent of the expanded coal production—is projected to enter world trade, with the largest investments to be made by producer-exporters. A comparable investment is needed for ships, and a relatively smaller amount of investment is needed for ports and internal transportation in importing countries.

3. The total investment over the period to the year 2000 is about $1,000 billion ($ 1978 U.S.) for WOCOL countries in the OECD whose aggregate gross capital forma- for the period is estimated to be about $38,000 billion. There are wide differences among countries, but spread out over the period of two decades, the amount of capital needed lies well within the potential combined capacity of domestic and international capital markets.

4. The large investments for mines and transport will not be made unless users—principally electric utilities and industrial organizations—make early decisions to build coal-using facilities and accompany such decisions by arrangements to secure their coal supplies. Such decisions will be necessary to ensure the financing of coal supply chains.

APPENDIX 1

SUMMARY COAL AND ENERGY PROJECTIONS
FOR WOCOL COUNTRIES

The following 24 pages contain two-page summaries of the coal projections for Cases A and B, prepared by teams of WOCOL members from each of the 12 OECD countries participating in WOCOL. The underlying energy and economic assumptions used in the analysis of future coal production, use, and trade for each country are also provided. The worksheets were designed by WOCOL to ensure the encoding of comparable national data for the various participating countries, and to allow the aggregation of coal and energy projections from individual countries into regional and global totals. Chapter 2 of this report, "Analysis of World Coal Prospects," is based on the information provided in these country worksheets.

Volume 2, *Future Coal Prospects: Country and Regional Assessments* contains a detailed description of the coal projections and the full texts of the national reports for the 16 countries participating in the World Coal Study, including the People's Republic of China, India, Indonesia, and Poland as well as the 12 OECD countries in WOCOL.

Australia

I. Coal Use, Production, and Trade	1977	1985 A	1985 B	1990 A	1990 B	2000 A	2000 B	1977-2000 Avg. annual growth—%/yr. A	1977-2000 Avg. annual growth—%/yr. B
Coal use in major markets (mtce)									
Metallurgical	8.3	10.3	10.3	12.2	12.2	16.8	16.8	3.1	3.1
Electric	25.0	42.9	42.9	64.2	64.2	107.3	107.3	6.5	6.5
Industry	4.4	11.1	11.1	13.5	13.5	16.1	16.1	5.8	5.8
Synthetic Fuels	–	–	–	–	–	–	24.9	–	–
Residential/Commercial	0.3	0.3	0.3	0.3	0.3	0.3	0.3	0	0
Total coal use	38.0	64.6	64.6	90.2	90.2	140.5	165.4	5.9	6.6
Distribution of coal use by market sector (%)									
Metallurgical	22	16	16	14	14	12	10	—	—
Electric	66	67	67	71	71	77	65	—	—
Industry	12	17	17	15	15	11	10	—	—
Synthetic Fuels	–	–	–	–	–	–	15	—	—
Residential/Commercial	0	0	0	0	0	0	0	—	—
Total coal use	100%	100%	100%	100%	100%	100%	100%	—	—
Coal consumption/imports (mtce) **Consumption**									
Metallurgical	8.3	10	10	12	12	17	17	3.2	3.2
Steam	29.7	55	55	78	78	124	149	6.4	7.3
Total coal consumption	38.0	65	65	90	90	141	166	5.9	6.6
Imports									
Metallurgical									
Steam									
Total coal imports									
Coal production/exports (mtce) **Production**									
Metallurgical	42.6	71	71	80	80	102	102	3.9	3.9
Steam	33.1	72	72	115	115	199	224	8.1	8.7
Total coal production	75.7	143	143	195	195	301	326	6.2	6.6
Exports									
Metallurgical	34.3	61	61	68	68	85	85	4.0	4.0
Steam	3.4	17	17	37	37	75	75	14.4	14.4
Total coal export	37.7	78	78	105	105	160	160	6.5	6.5

Australia

II. Coal's Role in Total Energy System	1977	1985		1990		2000		1977-2000 Avg. annual growth–%/yr.	
		A	B	A	B	A	B	A	B
Total Primary Energy Use (mtce)									
Oil, Domestic[1]	31.3	32.1	36.0	23.3	44.8	9.0	41.6	-5.3	1.2
Oil, Imported	14.3	21.0	14.2	32.9	5.0	62.7	6.3	6.6	-3.5
Gas, Domestic	9.4	16.0	16.0	22.9	22.9	35.4	35.4	6.0	6.0
Gas, Imported	--	--	--	--	--	--	--	--	--
Nuclear	--	--	--	--	--	--	--	--	--
Hydro, Solar, Other	9.3	10.2	10.2	10.7	10.7	11.5	11.5	0.9	0.9
Coal, Domestic	38.0	64.6	64.6	90.2	90.2	140.5	165.4	5.9	6.6
Coal, Imported	--	--	--	--	--	--	--	--	--
Total energy use	102.3	144	141	180	174	259	260	4.1	4.1
Coal penetration (%)	37	45	46	50	52	54	64	—	—
Total primary energy (mtce) input to electricity									
Oil and Gas	2.6	4.5	4.5	4.3	4.3	3.9	3.9	1.8	1.8
Hydro, Solar, Other	6.1	7.0	7.0	7.4	7.4	8.1	8.1	1.2	1.2
Nuclear	--	--	--	--	--	--	--	--	--
Coal	25.0	43.9	42.9	64.2	64.2	107	107	6.5	6.5
Total energy input	33.7	54.4	54.4	75.9	75.9	119	119	5.7	5.7
Coal penetration (%)	74	79	79	85	85	90	90	—	—
Total electric capacity (GWe)									
Oil and Gas	1.7	2.5	2.5	2.5	2.5	2.3	2.3	1.3	1.3
Hydro, Solar, Other	5.7	6.6	6.6	7.0	7.0	7.7	7.7	1.4	1.4
Nuclear	--	--	--	--	--	--	--	--	--
Coal	13.4	24.9	24.9	36.5	36.5	62	62	6.9	6.9
Total capacity	20.8	34	34	46	46	72	72	5.5	5.5
Coal Penetration (%)	64.4	73	73	79	79	86	86	1.3	1.3
Peak load	15	25	25	34	34	54	54	5.7	5.7
Peak reserve margin (%)	28	27	27	26	26	25	25	—	—
Total oil imports (mbd)	0.2	0.3	0.2	0.4	0.1	0.8	0.1	6.6	-3.5
Total oil consumption (mbd)									
Transportation	0.35	0.44	0.41	0.47	0.42	0.62	0.54	2.6	2.0
Residential/Commercial	0.05	0.05	0.05	0.06	0.05	0.07	0.06	1.5	0.9
Industry—Boilers	0.04	0.03	0.03	0.03	0.03	0.03	0.03	-1.3	-2.2
Industry—Other	0.16	0.17	0.16	0.18	0.16	0.23	0.20	1.5	0.9
Electric utilities	0.01	0.02	0.02	0.01	0.01	0.01	0.01	-0.5	-0.5
Total oil consumption	0.61	0.71	0.67	0.75	0.67	0.96	0.84	1.8	1.2
World oil price assumed for national coal analysis (1979 U.S. dollars/barrel)	$20*	20	35	20	40	25	50	1.0	4.1
Economic growth assumed for national coal analysis (GNP, billion 1978 dollars)	95.6	126	126	149	149	210	210	3.5	3.5

(1) Includes oil from shale but excludes oil from coal (which is counted as coal).

223

Canada

I. Coal Use, Production, and Trade	1977	1985		1990		2000		1977-2000 Avg. annual growth—%/yr.	
		A	B	A	B	A	B	A	B
Coal use in major markets (mtce)									
Metallurgical	7	9	9	11	11	15	15	–	–
Electric (includes gasification	17	31	28	37	40	50	65	4.8	6.0
Industry in industry)	1	4	4	7	10	11	20	11.0	14.6
Synthetic Fuels (includes use in oil sands)	–	–	–	–	11	6	21	–	–
Residential/Commercial	small	–	–	–	–	–	–	–	–
Total coal use	25	44	41	55	72	82	121	5.3	7.1
Distribution of coal use by market sector (%)									
Metallurgical	26	21	21	20	15	18	12	—	—
Electric	67	70	68	67	55	61	54	—	—
Industry	7	9	11	13	15	13	17	—	—
Synthetic Fuels	–	–	–	–	15	8	17	—	—
Residential/Commercial	small	neg.	neg.	neg.	neg.	neg.	neg.	—	—
Total coal use	100%	100%	100%	100%	100%	100%	100%	—	—
Coal consumption/imports (mtce) Consumption									
Metallurgical	7	9	9	11	11	15	15	3.4	3.4
Steam	18	35	32	44	61	67	106	5.9	8.0
Total coal consumption	25	44	41	55	72	82	121	5.3	7.1
Imports									
Metallurgical	7	7	5	8	5	9	5	–	–
Steam	6	6	5	7	5	8	4	–	–
Total coal imports	13	13	10	15	10	17	9	–	–
Coal production/exports (mtce) Production									
Metallurgical	10	15	17	18	21	29	33	4.7	5.3
Steam	13	33	31	41	66	63	126	7.1	10.4
Total coal production	23	48	48	59	87	92	159	6.2	8.8
Exports									
Metallurgical	10	13	13	15	15	23	23	3.7	3.7
Steam	1	4	4	4	10	4	24	6.2	14.8
Total coal export	11	17	17	19	25	27	47	4.0	6.5

Canada

II. Coal's Role in Total Energy System	1977	1985 A	1985 B	1990 A	1990 B	2000 A	2000 B	1977-2000 Avg. annual growth–%/yr. A	1977-2000 Avg. annual growth–%/yr. B
Total Primary Energy Use (mtce)									
Oil, Domestic	109	97	101	95	109	106	127	--	--
Oil, Imported	23	46	46	46	46	46	46	--	--
Gas, Domestic	56	67	90	72	92	82	93	1.6	2.2
Gas, Imported	(36)	--	--	--	--	--	--	--	--
Nuclear	9	21	36	27	53	45	88	7.2	10.4
Hydro, Solar, Other	75	84	82	111	94	148	117	3.1	2.0
Coal, Domestic	12	31	31	40	62	65	112	7.6	10.2
Coal, Imported	13	13	10	15	10	17	9	--	--
Total energy use	297	359	396	406	466	509	592	2.4	3.0
Coal penetration (%)	9	12	10	14	15	16	20	—	—
Total primary energy (mtce) input to electricity									
Oil and Gas	8	7	7	5	5	5	3	--	--
(1)Hydro, Solar, Other	74	95	83	111	95	145	118	3.0	2.1
Nuclear	9	21	36	27	53	45	88	7.2	10.4
Coal	17	31	28	37	40	50	65	4.8	6.0
Total energy input	108	154	154	180	193	245	274	3.6	4.1
Coal penetration (%)	15	20	18	21	21	20	24	—	—
Total electric capacity (GWe)									
Oil and Gas	10	10	10	9	9	7	7	--	--
Hydro, Solar, Other	43	58	58	66	66	77	75	2.6	2.5
Nuclear	5	10	15	13	25	26	67	7.4	11.9
Coal	12	20	20	22	30	38	61	5.2	7.3
Total capacity	70	98	103	110	130	148	210	3.3	4.9
Coal Penetration (%)	17	20	19	20	23	26	29	1.9	2.3
Peak load	52	74	75	83	96	111	158	3.3	4.9
Peak reserve margin (%)	25	25	25	25	25	25	25	—	—
Total oil imports (mbd)	0.3	0.6	0.6	0.6	0.6	0.6	0.6	--	--
Total oil consumption (mbd)									
Transportation	0.8	0.9	0.9	0.9	1.0	0.9	1.1	--	--
Residential/Commercial	0.3	0.3	0.3	0.3	0.3	0.4	0.4	--	--
Industry—Boilers	0.3	0.3	0.3	0.3	0.4	0.3	0.4	--	--
Industry—Other	0.2	0.3	0.3	0.3	0.3	0.4	0.4	--	--
Electric utilities	0.1	0.1	0.1	--	--	--	--	--	--
Total oil consumption	1.7	1.9	1.9	1.8	2.0	2.0	2.3	--	--
World oil price assumed for national coal analysis (1979 U.S. dollars/barrel)	$20*	20	20	23	26	29	42	2.5% after '83	5.0% after '83
Economic growth assumed for national coal analysis (GNP, billion 1978 dollars)	197	259	259	308	308	434	434	3.5	3.5

(1) Hydro, Solar & Other values calculated by difference. Discrepancies with respect to Total Primary Energy Use are probably due to net electrical exports not otherwise accounted for.

225

Denmark

I. Coal Use, Production, and Trade	1977	1985		1990		2000		1977-2000 Avg. annual growth—%/yr.	
		A	B	A	B	A	B	A	B
Coal use in major markets (mtce)									
Metallurgical	0	0	0	0	0	0	0	0	0
Electric	3.8	9.9	9.9	12.9	12.9	8.6	18.9	3.6	7.2
Industry	0.7	0.7	1.0	0.7	1.1	0.7	1.4	0	3.1
Synthetic Fuels	0	0	0	0	0	0	0	–	–
Residential/Commercial	0.1	0.1	0.2	0.1	0.4	0.1	0.6	0	8.1
Total coal use	4.6	10.7	11.1	13.7	14.4	9.4	20.9	3.2	6.8
Distribution of coal use by market sector (%)									
Metallurgical	0	0	0	0	0	0	0	——	——
Electric	83	93	89	94	90	91	90	——	——
Industry	15	6	9	5	7	8	7	——	——
Synthetic Fuels	0	0	0	0	0	0	0	——	——
Residential/Commercial	2	1	2	1	3	1	3	——	——
Total coal use	100%	100%	100%	100%	100%	100%	100%	——	——
Coal consumption/imports (mtce) Consumption									
Metallurgical	0	0	0	0	0	0	0	0	0
Steam	4.6	10.7	11.1	13.7	14.4	9.4	20.9	3.6	6.8
Total coal consumption	4.6	10.7	11.1	13.7	14.4	9.4	20.9	3.6	6.8
Imports									
Metallurgical	0	0	0	0	0	0	0	0	0
Steam	4.6	10.7	11.1	13.7	14.4	9.4	20.9	3.6	6.8
Total coal imports	4.6	10.7	11.1	13.7	14.4	9.4	20.9	3.6	6.8
Coal production/exports (mtce) Production (Greenland)									
Metallurgical	0	0	0	0	0	0	0	–	–
Steam	0	0	0.6	0	0.6	0	0.6	–	–
Total coal production	0	0	0.6	0	0.6	0	0.6	–	–
Exports (Greenland)									
Metallurgical	0	0	0	0	0	0	0	–	–
Steam	0	0	0.3	0	0.3	0	0.3	–	–
Total coal export	0	0	0.3	0	0.3	0	0.3	–	–

Denmark

II. Coal's Role in Total Energy System	1977	1985 A	1985 B	1990 A	1990 B	2000 A	2000 B	1977-2000 Avg. annual growth–%/yr. A	1977-2000 Avg. annual growth–%/yr. B
Total Primary Energy Use (mtce)									
Oil, Domestic	0.7	1.0	1.2	1.3	2.0	1.4	2.0	3.1	4.7
Oil, Imported	22.9	20.8	20.2	20.2	15.8	20.0	13.0	-0.6	-2.4
Gas, Domestic	0	2.4	2.4	4.7	4.7	4.7	4.7	--	--
Gas, Imported	0	0	0	0	0	0	0	--	--
Nuclear	0	0	0	0	0	13.2	0	--	--
Hydro, Solar, Other	0	0	0	0	0	0	0	--	--
Coal, Domestic	0	0	0	0 .	0	0	0	--	--
Coal, Imported	4.6	10.7	11.1	13.7	14.4	9.4	20.9	3.2	6.8
Total energy use	28.2	34.9	34.9	39.9	36.9	48.7	40.6	2.4	1.6
Coal penetration (%)	16	31	32	34	39	19	51	—	—
Total primary energy (mtce) input to electricity									
Oil and Gas	3.8	1.7	1.7	1.5	1.5	1.7	1.7	-3.4	-3.4
Hydro, Solar, Other	0	0	0	0	0	0	0	--	--
Nuclear	0	0	0	0	0	13.2	0	--	--
Coal	3.8	9.9	9.9	12.9	12.9	8.6	18.9	3.6	7.2
Total energy input	7.6	11.6	11.6	14.4	14.4	23.5	20.6	5.0	4.4
Coal penetration (%)	50	85	85	90	90	37	92	—	—
Total electric capacity (GWe)									
Oil and Gas	3.3	1.9	1.9	1.9	1.9	1.9	1.9	-2.4	-2.4
Hydro, Solar, Other	0	0	0	0	0	0	0	--	--
Nuclear	0	0	0	0	0	6.5	0	--	--
Coal	3.0	6.6	6.6	8.5	8.5	8.5	13.0	4.6	6.6
Total capacity	6.3	8.5	8.5	10.4	10.4	16.9	14.9	4.4	3.8
Coal Penetration (%)	48	78	78	82	82	50	87	—	—
Peak load	4.3	6.8	6.8	8.7	8.7	14.1	12.4	5.3	4.7
Peak reserve margin (%)	45	20	20	20	20	20	20	—	—
Total oil imports (mbd)	0.30	0.27	0.27	0.27	0.21	0.26	0.17	-0.6	-2.4
Total oil consumption (mbd)									
Transportation	0.07								
Residential/Commercial	0.13								
Industry—Boilers	0.06								
Industry—Other									
Electric utilities	0.05	0.02	0.02	0.02	0.02	0.02	0.02	-3.9	-3.9
Total oil consumption	0.31	0.29	0.28	0.28	0.23	0.30	0.20	-0.01	-1.9
World oil price assumed for national coal analysis (1979 U.S. dollars/barrel)	$20*	20	20	20	25	30	40	1.8	3.1
Economic growth assumed for national coal analysis (GNP, billion 1978 dollars)	52.1	69.1	69.1	79.6	75.5	112.4	89.9	3.4	2.4

* Uses current price of $20/barrel (June 1979 U.S. dollars) as baseline world oil price and as floor price throughout the period.

227

Finland

I. Coal Use, Production, and Trade	1977	1985 A	1985 B	1990 A	1990 B	2000 A	2000 B	1977-2000 Avg. annual growth—%/yr. A	1977-2000 Avg. annual growth—%/yr. B
Coal use in major markets (mtce)									
Metallurgical	0.8	1.0	1.0	1.0	1.0	1.0	1.0	1.0	1.0
Electric	1.9	1.1	1.1	3.2	3.2	3.2	7.9	2.3	6.4
Industry	0.7	0.9	0.9	1.0	1.0	1.0	1.0	1.6	1.6
Synthetic Fuels	–	–	–	–	–	–	–	–	–
Residential/Commercial [1]	0.9	1.4	1.4	2.3	2.3	3.5	3.5	6.1	6.1
Total coal use	4.3	4.4	4.4	7.5	7.5	8.7	13.4	3.1	5.1
Distribution of coal use by market sector (%)									
Metallurgical	19	23	23	13	13	11	7	—	—
Electric	44	25	25	43	43	37	59	—	—
Industry	16	20	20	13	13	12	8	—	—
Synthetic Fuels	–	–	–	–	–	–	–	—	—
Residential/Commercial [1]	21	32	32	31	31	40	26	—	—
Total coal use	100%	100%	100%	100%	100%	100%	100%	—	—
Coal consumption/imports (mtce) Consumption									
Metallurgical	0.8	1.0	1.0	1.0	1.0	1.0	1.0	1.0	1.0
Steam	3.5	3.4	3.4	6.5	6.5	7.7	12.4	3.5	5.7
Total coal consumption	4.3	4.4	4.4	7.5	7.5	8.7	13.4	3.1	5.1
Imports									
Metallurgical	0.9	1.0	1.0	1.0	1.0	1.0	1.0	0.5	0.5
Steam	4.1	3.4	3.4	6.5	6.5	7.7	12.4	2.8	4.9
Total coal imports	5.0	4.4	4.4	7.5	7.5	8.7	13.4	2.4	4.4
Coal production/exports (mtce) Production									
Metallurgical	–	–	–	–	–	–	–		
Steam	–	–	–	–	–	–	–		
Total coal production	–	–	–	–	–	–	–		
Exports									
Metallurgical	–	–	–	–	–	–	–		
Steam	–	–	–	–	–	–	–		
Total coal export	–	–	–	–	–	–	–		

(1) Includes power generation in district heating plants.

228

Finland

II. Coal's Role in Total Energy System	1977	1985		1990		2000		1977-2000 Avg. annual growth-%/yr.	
		A	B	A	B	A	B	A	B
Total Primary Energy Use (mtce)									
Oil, Domestic	--	--	--	--	--	--	--	--	--
Oil, Imported	17.1	18.5	18.5	17.0	17.0	15.5	15.5	-0.4	-0.4
Gas, Domestic	0.3	0.4	0.4	0.4	0.4	0.4	0.4	1.3	1.3
Gas, Imported	1.1	1.3	1.3	1.4	1.4	1.6	1.6	1.6	1.6
Nuclear	0.9	5.0	5.0	5.0	5.0	9.7	5.0	10.9	7.7
Hydro, Solar, Other	9.2	11.4	11.4	12.7	12.7	15.1	15.1	2.2	2.2
Coal, Domestic	--	--	--	--	--	--	--		
Coal, Imported	3.5	3.4	3.4	6.5	6.5	7.7	12.4	3.5	5.7
Total energy use	32.1	40.0	40.0	43.0	43.0	50.0	50.0	2.0	2.0
Coal penetration (%)	11	9	9	15	15	15	25	—	—
Total primary energy (mtce) input to electricity									
Oil and Gas	1.8	1.5	1.5	1.4	1.4	1.3	1.3	-1.4	-1.4
Hydro, Solar, Other	5.1	6.5	6.5	6.2	6.2	5.7	5.7	0.5	0.5
Nuclear	0.9	5.0	5.0	5.0	5.0	9.7	5.0	10.9	7.7
Coal	2.2	1.5	1.5	3.9	3.9	4.2	8.9	2.5	6.3
Total energy input	10.0	14.5	14.5	16.5	16.5	20.9	20.9	3.3	3.3
Coal penetration (%)	22	10	10	24	24	20	43	—	—
Total electric capacity (GWe)									
Oil and Gas	2.9	2.6	2.6	2.4	2.4	2.2	2.2	-1.2	-1.2
Hydro, Solar, Other	2.9	4.8	4.8	4.9	4.9	5.6	5.6	2.9	2.9
Nuclear	0.4	2.2	2.2	2.2	2.2	4.2	2.2	10,8	7.7
Coal	1.9	2.4	2.4	3.0	3.0	3.5	5.5	2.7	4.7
Total capacity	8.1	12.0	12.0	12.5	12.5	15.5	15.5	2.9	2.9
Coal Penetration (%)	23	20	20	24	24	23	35	—	—
Peak load	5.9	9.8	9.8	10.2	10.2	12.6	12.6	3.4	3.4
Peak reserve margin (%)	37	23	23	23	23	23	23	—	—
Total oil imports (mbd)	0.29	0.26	0.26	0.24	0.24	0.22	0.22	-1.2	-1.2
Total oil consumption (mbd)									
Transportation	0.06	0.07	0.07	0.08	0.08	0.08	0.08		
Residential/Commercial	0.09	0.08	0.08	0.07	0.07	0.06	0.06		
Industry—Boilers	0.07	0.09	0.09	0.08	0.08	0.07	0.07		
Industry—Other									
Electric utilities	0.02	0.02	0.02	0.01	0.01	0.01	0.01		
Total oil consumption	0.24	0.26	0.26	0.24	0.24	0.22	0.22	-0.4	-0.4
World oil price assumed for national coal analysis (1979 U.S. dollars/barrel)	$20*	25	25	30	30	40	40	3.1	3.1
Economic growth assumed for national coal analysis (GNP, billion 1978 dollars)	30	38	38	44	44	59	59	3.0	3.0

* Uses current price of $20/barrel (June 1979 U.S. dollars) as baseline world oil price and as floor price throughout the period.

France

I. Coal Use, Production, and Trade	1977	1985		1990		2000		1977-2000 Avg. annual growth—%/yr.	
		A	B	A	B	A	B	A	B
Coal use in major markets (mtce)									
Metallurgical	14	16	17	16	18	17	20	0.8	1.4
Electric	22	12	25	10	30	10	45	-3.5	3.2
Industry	3	5	12	7	19	14	40	6.9	11.9
Synthetic Fuels	–	–	–	–	–	3	10	–	–
Residential/Commercial	6	2	5	2	7	4	10	-1.6	2.1
Total coal use	45	35	59	35	74	48	125	0.3	4.5
Distribution of coal use by market sector (%)									
Metallurgical	31	46	29	46	24	36	16	—	—
Electric	49	34	42	28	11	21	36	—	—
Industry	7	14	20	20	26	29	32	—	—
Synthetic Fuels	–	–	–	–	–	6	8	—	—
Residential/Commercial	13	6	9	6	9	8	8	—	—
Total coal use	100%	100%	100%	100%	100%	100%	100%	—	—
Coal consumption/imports (mtce) **Consumption**									
Metallurgical	14	16	17	16	18	17	20	0.8	1.6
Steam	31	19	42	19	56	31	105	–	5.4
Total coal consumption	45	35	59	35	74	48	125	0.3	4.5
Imports									
Metallurgical	10	11	12	11	13	12	15	0.8	1.8
Steam	14	11	34	14	51	26	100	2.7	8.9
Total coal imports	24	22	46	25	64	38	115	2.0	7.0
Coal production/exports (mtce) **Production**									
Metallurgical	8	5	5	5	5	5	5	-2.0	-2.0
Steam	13	8	8	5	5	5	5	-4.2	-4.2
Total coal production	21	13	13	10	10	10	10	-3.3	-3.3
Exports									
Metallurgical	2*	–	–	–	–	–	–	–	–
Steam	–	–	–	–	–	–	–	–	–
Total coal export	2	–	–	–	–	–	–	–	–

* Coke in terms of coking coal.

France

II. Coal's Role in Total Energy System	1977	1985 A	1985 B	1990 A	1990 B	2000 A	2000 B	1977-2000 Avg. annual growth–%/yr. A	1977-2000 Avg. annual growth–%/yr. B
Total Primary Energy Use (mtce)									
Oil, Domestic	1	1	1	1	1	1	1		
Oil, Imported	149	136	136	135	135	147	134		
Gas, Domestic	9	5	5	5	5	5	5		
Gas, Imported	20	44	47	52	57	62	73		
Nuclear	5	62	58	90	85	137	140		
Hydro, Solar, Other	26	24	24	25	28	33	45		
Coal, Domestic	21	13	13	10	10	10	10		
Coal, Imported	24	22	46	25	64	38	115		
Total energy use	255	307	330	343	385	430	520		
Coal penetration (%)	18	11	18	10	19	11	24	—	—
Total primary energy (mtce) input to electricity									
Oil and Gas	20.4	13	14	12	10	10	7		
Hydro, Solar, Other	25.3	21	21	21	23	23	25		
Nuclear	5.7	62	58	90	85	137	140		
Coal	22.0	12	25	10	30	10	45		
Total energy input	73.7	108	118	133	148	180	217		
Coal penetration (%)	29	11	21	8	20	6	21	—	—
Total electric capacity (GWe)									
Oil and Gas	18.9	15	17	15	12	14	9		
Hydro, Solar, Other	17.5	20	20	22	23	25	31		
Nuclear	4.6	36	34	48	50	78	80		
Coal	10.3	14	16	10	20	13	30		
Total capacity	51.3	85	87	95	105	130	150		
Coal Penetration (%)	20	16	18	11	19	10	20	—	—
Peak load	37								
Peak reserve margin (%)	38							—	—
Total oil imports (mbd)	2.1	1.9	1.9	1.9	1.9	2.1	1.9	–	-0.4
Total oil consumption (mbd)									
Transportation	0.66								
Residential/Commercial	0.56								
Industry—Boilers)	0.69								
Industry—Other									
Electric utilities	0.19								
Total oil consumption	2.1								
World oil price assumed for national coal analysis (1979 U.S. dollars/barrel)	$20*	20	25	25	30	30	40	3.2	4.5
Economic growth assumed for national coal analysis (GNP, billion 1978 dollars)	381	483	521	560	634	756	939	3.0	4.0

* Uses current price of $20/barrel (June 1979 U.S. dollars) as baseline world oil price and as floor price throughout the period.

Germany, Federal Republic of

I. Coal Use, Production, and Trade	1977	1985		1990		2000		1977-2000 Avg. annual growth—%/yr.	
		A	B	A	B	A	B	A	B
Coal use in major markets (mtce)									
Metallurgical	23	26	24	25	23	25	22	0.4	0.2
Electric	61	76	79	84	89	99	106	2.1	2.4
Industry	5	6	8	6	9	8	12	2.1	3.9
Synthetic Fuels	–	1	3	5	10	10	25	–	–
Residential/Commercial	13	10	12	9	11	8	10	-2.1	-1.1
Total coal use	102	119	126	129	142	150	175	1.7	2.4
Distribution of coal use by market sector (%)									
Metallurgical	22	22	19	19	16	17	13	—	—
Electric	60	64	63	65	63	66	61	—	—
Industry	5	5	6	5	6	5	7	—	—
Synthetic Fuels	–	1	2	4	7	7	14	—	—
Residential/Commercial	13	8	10	7	8	5	5	—	—
Total coal use	100%	100%	100%	100%	100%	100%	100%	—	—
Coal consumption/imports (mtce) **Consumption**									
Metallurgical	23	26	24	25	23	25	22	0.4	-0.2
Steam	79	93	102	104	119	125	153	2.0	2.9
Total coal consumption	102	119	126	129	142	150	175	1.7	2.4
Imports									
Metallurgical	1	–	–	–	–	–	–	–	–
Steam	8	9	11	18	26	20	40	4.1	7.2
Total coal imports	9	9	11	18	26	20	40	3.5	6.7
Coal production/exports (mtce) **Production**									
Metallurgical	40	44	37	45	38	45	37	0.5	-0.3
Steam	80	84	91	86	93	105	113	1.2	1.5
Total coal production	120	128	128	131	131	150	150	1.0	1.0
Exports									
Metallurgical	17	18	13	20	15	20	15	0.7	-0.5
Steam	5	–	–	–	–	–	–	–	–
Total coal export	22	18	13	20	15	20	15	-0.4	-1.7

Germany, Federal Republic of

II. Coal's Role in Total Energy System	1977	1985 A	1985 B	1990 A	1990 B	2000 A	2000 B	1977-2000 Avg. annual growth–%/yr. A	B
Total Primary Energy Use (mtce)									
Oil, Domestic	194	223	212	223	205	160	130	-0.8	-1.7
Oil, Imported									
Gas, Domestic	55	88	85	92	87	95	90	2.4	2.2
Gas, Imported									
Nuclear	12	37	34	55	50	125	105	10.7	9.9
Hydro, Solar, Other	9	11	11	16	16	40	40	6.7	6.7
Coal, Domestic	93	113	115	111	116	130	135		
Coal, Imported	9	9	11	18	26	20	40	3.5	6.7
Total energy use	372	478	468	515	500	570	540	1.9	1.6
Coal penetration (%)	27.4	24.9	26.9	25.0	28.4	26.3	32.4	—	—
Total primary energy (mtce) input to electricity									
Oil and Gas	25	33	29	31	24	13	10	-5.2	-7.3
Hydro, Solar, Other	12	17	17	17	17	19	19	3.6	3.6
Nuclear	12	37	34	55	50	125	105	10.7	9.9
Coal	61	76	79	84	89	99	106	2.1	2.4
Total energy input	110	163	159	187	180	256	240	3.7	3.5
Coal penetration (%)	55.5	46.6	49.7	44.9	49.4	38.7	44.2	—	—
Total electric capacity (GWe)									
Oil and Gas	36.3	39	36	37	33	32.5	30	-0.9	-1.5
Hydro, Solar, Other									
Nuclear	7.4	19	17	28	25	62.5	53	9.7	8.9
Coal	40.0	50	52	55	57	65	67	2.1	2.3
Total capacity	83.7	108	105	120	115	160	150	2.9	2.6
Coal Penetration (%)								—	—
Peak load									
Peak reserve margin (%)								—	—
Total oil imports (mbd)	2.58	3.0	2.84	3.02	2.77	2.16	1.75	-0.8	-1.7
Total oil consumption (mbd)									
Transportation	0.69	0.75	0.74	0.81	0.79	0.72	0.69	0.2	—
Residential/Commercial	0.90	1.03	1.00	0.94	0.86	0.62	0.46	-1.6	-2.9
Industry—Boilers	0.70	0.86	0.79	0.88	0.82	0.62	0.39	-0.5	-2.5
Industry—Other									
Electric utilities	0.11	0.13	0.10	0.13	0.08	0.04	0.04	-4.3	-4.3
Total oil consumption	2.69	3.10	2.94	3.10	2.85	2.22	1.81	-0.8	-1.7
World oil price assumed for national coal analysis (1979 U.S. dollars/barrel)	$20*	21.3	23.0	22.2	25.2	24	30	0.8	1.0
Economic growth assumed for national coal analysis (GNP, billion 1978 dollars)	514.1	667	641	784	736	1085	970	3.3	2.8

* Uses current price of $20/barrel (June 1979 U.S. dollars) as baseline world oil price and as floor price throughout the period.

Italy

I. Coal Use, Production, and Trade	1977	1985 A	1985 B	1990 A	1990 B	2000 A	2000 B	1977-2000 Avg. annual growth—%/yr. A	B
Coal use in major markets (mtce)									
Metallurgical	11.1	11.0	11.0	11.5	11.5	12.0	12.0	0.3	0.3
Electric	1.8	10.0	10.0	9.3	18.8	17.0	37.5	–	–
Industry	0.2	1.0	1.5	2.0	2.0	2.5	3.0	–	–
Synthetic Fuels	–	–	–	–	–	–	8.0	–	–
Residential/Commercial	0.4	0.3	0.4	0.2	0.2	–	–	–	–
Total coal use	13.5	22.3	22.9	23.0	32.5	31.5	60.5	3.7	6.7
Distribution of coal use by market sector (%)									
Metallurgical	82.2	49.3	48.0	50.0	35.4	38.1	19.8	—	—
Electric	13.3	44.9	43.7	40.4	57.8	54.0	62.0	—	—
Industry	1.5	4.5	6.6	8.7	6.2	7.9	5.0	—	—
Synthetic Fuels	–	–	–	–	–	–	13.2	—	—
Residential/Commercial	3.0	1.3	1.7	0.9	0.6	–	–	—	—
Total coal use	100%	100%	100%	100%	100%	100%	100%	—	—
Coal consumption/imports (mtce) **Consumption**									
Metallurgical	11.1	11.0	11.0	11.5	11.5	12.0	12.0	0.3	0.3
Steam	2.4	11.3	11.9	11.5	21.0	19.5	48.5	9.5	–
Total coal consumption	13.5	22.3	22.9	23.0	32.5	31.5	60.5	3.7	6.7
Imports									
Metallurgical	11.1	11.0	11.0	11.5	11.5	12.0	12.0	0.3	0.3
Steam	2.0	10.3	10.9	10.5	20.0	16.5	45.5	9.6	–
Total coal imports	13.1	21.3	21.9	22.0	31.5	28.5	57.5	3.4	6.6
Coal production/exports (mtce) **Production**									
Metallurgical									
Steam	0.4	1.0	1.0	1.0	1.0	3.0	3.0	9.1	9.1
Total coal production	0.4	1.0	1.0	1.0	1.0	3.0	3.0	9.1	9.1
Exports									
Metallurgical									
Steam									
Total coal export									

Italy

II. Coal's Role in Total Energy System	1977	1985 A	1985 B	1990 A	1990 B	2000 A	2000 B	1977-2000 Avg. annual growth–%/yr. A	1977-2000 Avg. annual growth–%/yr. B
Total Primary Energy Use (mtce)									
Oil, Domestic	1.6	3.6	3.6	4.3	4.3	5.5	5.5	1.8	1.8
Oil, Imported	135.5	158.1	155.5	160.2	148.7	156.5	145.5	0.3	0.3
Gas, Domestic	14.3	14.0	14.0	14.0	14.0	14.0	14.0	–	–
Gas, Imported	16.7	30.5	30.5	49.0	49.0	63.0	63.0	5.9	5.9
Nuclear	1.0	2.5	2.5	9.5	9.5	36.5	24.5	–	–
Hydro, Solar, Other	18.1	21.0	21.0	26.0	26.0	35.0	39.0	2.9	3.4
Coal, Domestic	0.4	1.0	1.0	1.0	1.0	3.0	3.0	9.1	9.1
Coal, Imported	13.1	21.3	21.9	22.0	31.5	28.5	57.5	3.4	6.6
Total energy use	200.7	252.0	250.0	286.0	284.0	342.0	352.0	2.3	2.4
Coal penetration (%)	6.7	8.8	9.2	8.0	11.4	9.2	17.2	—	—
Total primary energy (mtce) input to electricity									
Oil and Gas	31.5	44.5	44.5	46.2	44.7	33.5	34.0	0.3	0.3
Hydro, Solar, Other	18.1	19.0	19.0	20.0	20.0	21.0	22.0	0.6	0.8
Nuclear	1.0	2.5	2.5	9.5	9.5	36.5	24.5	–	–
Coal	1.8	10.0	10.0	9.3	18.8	17.0	37.5	–	–
Total energy input	52.4	76.0	76.0	85.0	93.0	108.0	118.0	3.2	3.6
Coal penetration (%)	3.4	13.1	13.1	10.9	20.2	15.7	31.8	—	—
Total electric capacity (GWe)									
Oil and Gas	18.3	30.2	30.2	32.3	32.3	28.0	28.0	2.8	1.8
Hydro, Solar, Other	14.4	16.5	16.5	17.5	17.5	18.0	20.0	0.9	1.4
Nuclear	0.5	1.5	1.5	5.5	5.5	21.0	14.0	–	–
Coal	2.4	6.6	6.6	6.0	11.7	10.8	23.0	6.7	–
Total capacity	35.6	54.8	54.8	61.3	67.0	77.8	85.0	3.4	3.8
Coal Penetration (%)	6.7	12.0	12.0	9.8	17.5	13.9	27.0	—	—
Peak load	28.3	45.7	45.7	51.0	55.8	64.8	71.0	3.7	4.1
Peak reserve margin (%)	20.5	20.0	20.0	20.0	20.0	20.0	20.0	—	—
Total oil imports (mbd)	1.89	2.21	2.17	2.24	2.08	2.19	2.04	0.6	0.3
Total oil consumption (mbd)									
Transportation	0.61	0.69	0.69	0.71	0.67	0.91	0.87	1.7	1.5
Residential/Commercial	0.45	0.45	0.44	0.44	0.38	0.37	0.29	-0.8	-1.9
Industry—Boilers	0.33	0.39	0.36	0.37	0.32	0.26	0.20	-1.0	-2.1
Industry—Other	0.13	0.17	0.18	0.23	0.24	0.32	0.35	4.0	4.4
Electric utilities	0.40	0.56	0.56	0.55	0.53	0.40	0.40	–	–
Total oil consumption	1.92	2.26	2.23	2.30	2.14	2.26	2.11	0.7	0.4
World oil price assumed for national coal analysis (1979 U.S. dollars/barrel)	$20*					25	40		
Economic growth assumed for national coal analysis (GNP, billion 1978 dollars)	223.8	287.9	287.9	330.5	341.2	423.0	472.1	2.8	3.3

* Uses current price of $20/barrel (June 1979 U.S. dollars) as baseline world oil price and as floor price throughout the period.

Japan

I. Coal Use, Production, and Trade	1977	1985		1990		2000		1977-2000 Avg. annual growth—%/yr.	
		A	B	A	B	A	B	A	B
Coal use in major markets (mtce)									
Metallurgical	69	80	84	84	86	86	92	1.0	1.3
Electric	6	13	13	30	37	57	72	10.3	11.4
Industry	4	4	5	5	7	7	12	2.5	4.9
Synthetic Fuels	–	–	–	–	7	–	48	...	...
Residential/Commercial	–	–	–	–	–	–	–	...	...
Total coal use	79	97	102	119	137	150	224	2.8	4.6
Distribution of coal use by market sector (%)									
Metallurgical	87	82	82	71	63	57	41	—	—
Electric	8	13	13	25	27	38	32	—	—
Industry	5	5	5	4	5	5	5	—	—
Synthetic Fuels	–	–	–	–	5	–	22	—	—
Residential/Commercial	–	–	–	–	–	–	–	—	—
Total coal use	100%	100%	100%	100%	100%	100%	100%	—	—
Coal consumption/imports (mtce) **Consumption**									
Metallurgical	69	80	84	84	86	86	92	1.0	1.3
Steam	10	17	18	35	44	64	84	8.1	9.7
Total coal consumption	79	97	102	119	130	150	176	2.8	4.6
Imports									
Metallurgical	60	73	76	77	79	79	85	1.2	1.5
Steam	2	6	7	24	33	53	73	15.3	16.9
Total coal imports	62	79	84	101	112	132	158	3.3	4.2
Coal production/exports (mtce) **Production**									
Metallurgical	8	7	7	7	7	7	7	...	...
Steam	9	11	11	11	11	11	11	...	...
Total coal production	17	18	18	18	18	18	18	...	...
Exports									
Metallurgical	–	–	–	–	–	–	–		
Steam	–	–	–	–	–	–	–		
Total coal export	–	–	–	–	–	–	–		

Note: Figures for "Total coal consumption" and "Total coal imports" exclude coal feedstock for synthetic liquid fuels imported into Japan. These estimates are included in "Synthetic Fuels" in "Coal use in major markets."

Japan

II. Coal's Role in Total Energy System	1977	1985		1990		2000		1977-2000 Avg. annual growth–%/yr.	
		A	B	A	B	A	B	A	B
Total Primary Energy Use (mtce)									
Oil, Domestic (incl. Gas)	5	11	11	12	13	19	19	6.0	6.0
Oil, Imported	398	471	476	483	490	566	542	1.5	1.3
Gas, Domestic									
Gas, Imported	15	48	51	73	74	80	83	7.5	7.7
Nuclear	11	50	54	92	98	188	199	13.1	13.4
Hydro, Solar, Other	26	27	28	29	33	31	61	0.8	3.8
Coal, Domestic	17	18	18	18	18	18	18	--	--
Coal, Imported	62	79	84	101	117**	132	189**	3.2	5.0
Total energy use	534	704	722	808	843	1,034	1,111	2.9	3.2
Coal penetration (%)	15	14	14	15	16	15	19	—	—
Total primary energy (mtce) input to electricity									
Oil and Gas	112	135	139	123	123	108	110	--	--
Hydro, Solar, Other	25	27	28	28	32	30	4	0.8	2.5
Nuclear	11	50	54	92	98	188	199	13.1	13.4
Coal	6	13	14	30	37	57	72	9.8	10.8
Total energy input	154	225	235	273	290	383	425	4.0	4.5
Coal penetration (%)	4	6	6	11	13	15	17	—	—
Total electric capacity (GWe)									
Oil and Gas	72	95	95	95	97	94	96	1.2	1.3
Hydro, Solar, Other	25	33	35	37	43	52	65	3.2	4.2
Nuclear	8	28	30	50	53	94	100	11.3	11.6
Coal	4	9	10	20	21	41	52	10.6	11.8
Total capacity	109	165	170	202	214	281	313	4.2	4.7
Coal Penetration (%)	4	5	6	10	10	15	17	—	—
Peak load	88	136	141	166	176	233	259	4.2	4.8
Peak reserve margin (%)	24	21	21	22	22	21	21	—	—
Total oil imports (mbd)	5.3	6.0	6.1	6.2	6.3	7.2	6.9	1.5	1.3
Total oil consumption (mbd)									
Transportation	1.1	1.5	1.6	1.6	1.8	1.9	1.9	2.4	2.4
Residential/Commercial	0.6	1.0	1.0	1.3	1.3	1.6	1.5	4.4	4.1
Industry—Boilers)	1.5	2.5	2.5	2.6	2.8	3.2	3.2	3.3	3.3
Industry—Other)									
Electric utilities	1.3	1.1	1.2	0.8	0.7	0.8	0.6	-2.1	-3.3
Total oil consumption	5.3	6.1	6.3	6.3	6.5	7.5	7.2	1.6	1.3
World oil price assumed for national coal analysis (1979 U.S. dollars/barrel)	$20*								
Economic growth assumed for national coal analysis (GNP, billion 1978 dollars)	912	1,258	1,307	1,538	1,636	2,298	2,566	4.1	4.6

* Uses current price of $20/barrel (June 1979 U.S. dollars) as baseline world oil price and as floor price throughout the period.

** Including liquefied coal (5 and 31 mtce respectively).

Netherlands

I. Coal Use, Production, and Trade	1977	1985		1990		2000		1977-2000 Avg. annual growth—%/yr.	
		A	B	A	B	A	B	A	B
Coal use in major markets (mtce)									
Metallurgical	3.0	3.4	3.4	3.3	3.6	2.9	4.0	–	–
Electric	1.1	3.9	3.9	6.0	8.6	13.7	25.9	10.4	13.5
Industry	0.3	2.2	2.2	2.3	2.9	3.0	4.0	10.5	11.9
Synthetic Fuels	–	–	–	–	–	1.3	2.4	–	–
Residential/Commercial	0.1	–	–	0.6	0.6	1.0	1.0	–	–
Total coal use	4.5	10.4	10.4	13.1	16.6	22.8	38.2	7.3	9.7
Distribution of coal use by market sector (%)									
Metallurgical	66.7	32.7	32.7	25.2	21.7	12.7	10.5	—	—
Electric	24.4	37.5	37.5	45.8	51.8	60.1	67.8	—	—
Industry	6.7	21.2	21.2	17.6	17.5	13.2	10.5	—	—
Synthetic Fuels	–	–	–	–	–	5.7	6.3	—	—
Residential/Commercial	2.2	–	–	4.6	3.6	4.4	2.6	—	—
Total coal use	100%	100%	100%	100%	100%	100%	100%	—	—
Coal consumption/imports (mtce) **Consumption**									
Metallurgical	3.0	3.4	3.4	3.3	3.6	2.9	4.0	–	1.3
Steam	1.5	7.0	7.0	9.8	13.0	19.9	34.2	10.4	14.5
Total coal consumption	4.5	10.4	10.4	13.1	16.6	22.8	38.2	7.3	9.7
Imports									
Metallurgical	3.0	3.4	3.4	3.3	3.6	2.9	4.0	–	1.3
Steam	1.5	7.0	7.0	9.8	13.0	19.9	34.2	10.4	14.5
Total coal imports	4.5	10.4	10.4	13.1	16.6	22.8	38.2	7.3	9.7
Coal production/exports (mtce) **Production**									
Metallurgical									
Steam									
Total coal production									
Exports									
Metallurgical									
Steam									
Total coal export									

Netherlands

II. Coal's Role in Total Energy System	1977	1985 A	1985 B	1990 A	1990 B	2000 A	2000 B	1977-2000 Avg. annual growth–%/yr. A	1977-2000 Avg. annual growth–%/yr. B
Total Primary Energy Use (mtce)									
Oil, Domestic	2.3	2.9	2.9	2.9	2.9	2.2	2.2	--	--
Oil, Imported	33.4	49.0	56.4	52.4	61.8	58.9	80.3	2.5	3.8
Gas, Domestic	46.6	38.8	38.8	38.1	38.2	32.8	32.8	--	--
Gas, Imported	--	9.2	9.2	10.1	10.1	10.8	10.8	--	--
Nuclear	1.4	1.4	1.4	1.4	1.4	8.0	1.4	7.9	--
Hydro, Solar, Other	--	0.3	0.3	0.3	0.3	2.5	2.5	--	--
Coal, Domestic	--	--	--	--	--	--	--	--	--
Coal, Imported	4.5	10.4	10.4	13.1	16.6	22.8	38.2	7.3	9.7
Total energy use	88.2	112.0	119.4	118.3	131.3	138.0	168.2	2.0	2.85
Coal penetration (%)	5.1	9.3	8.7	11.1	12.6	16.5	22.7	—	—
Total primary energy (mtce) input to electricity									
Oil and Gas	14.4	16.1	17.2	15.7	16.3	3.9	4.3	--	--
Hydro, Solar, Other	--	--	--	--	--	1.4	1.4	--	--
Nuclear	1.4	1.4	1.4	1.4	1.4	8.0	1.4		
Coal	1.4	3.9	3.9	6.0	8.6	13.7	25.9	10.4	13.5
Total energy input	17.2	21.4	22.5	23.1	26.3	27.0	33.0	2.0	2.9
Coal penetration (%)	8	18	17	26	33	51	78	—	—
Total electric capacity (GWe)									
Oil and Gas	13.3	12.8	12.8	13.0	12.0	8.0	6.0	--	--
Hydro, Solar, Other	--	--	--	--	--	( )	( )	--	--
Nuclear	0.5	0.5	0.5	0.5	0.5	5.0	0.5	10.5	--
Coal	0.7	2.2	2.2	4.0	5.5	8.5	16.5	11.3	14.6
Total capacity	14.5	15.5	15.5	17.5	18.0	21.5	23.0	1.7	2.0
Coal Penetration (%)	5	15	15	25	30	40	70	—	—
Peak load	8.7	11.3	11.3	12.8	13.1	15.7	16.8	2.6	2.9
Peak reserve margin (%)	65	27	27	27	27	27	27	—	—
Total oil imports (mbd)									
Total oil consumption (mbd)									
Transportation	0.2	0.2	0.2	0.2	0.2	0.2	0.3	--	--
Residential/Commercial		--	--	--	--	--	--	--	--
Industry—Boilers		0.1	0.1	0.1	0.1	0.1	0.1	--	--
Industry—Other	0.1	0.2	0.2	0.3	0.4	0.5	0.7	--	--
Electric utilities	0.1	0.1	0.1	0.1	0.1	--	--	--	--
Total oil consumption	0.4	0.6	0.8	0.7	0.8	0.8	1.1	3.1	4.5
World oil price assumed for national coal analysis (1979 U.S. dollars/barrel)	$20*					30	25	1.8	1.0
Economic growth assumed for national coal analysis (GNP, billion 1978 dollars)	120	146	152	161	187	198	237	2.2	3.0

* Uses current price of $20/barrel (June 1979 U.S. dollars) as baseline world oil price and as floor price throughout the period.

Sweden

I. Coal Use, Production, and Trade	1977	1985		1990		2000		1977-2000 Avg. annual growth—%/yr.	
		A	B	A	B	A	B	A	B
Coal use in major markets (mtce)									
Metallurgical	1.8	2.2	2.2	2.4	2.4	2.8	2.8	1.9 [1)	1.9 [1)
Electric	0	0.4	0.7	0.8	6.3	6.7	15.2	20.7	22.8
Industry	0.3	1.8	1.8	2.8	2.8	3.3	3.3	11.0	11.0
Synthetic Fuels	–	–	–	–	–	0.7	0.7	– [1)	– [1)
Residential/Commercial	0	0.7	0.7	1.5	2.4	3.6	3.9	7.4	8.1
Total coal use	2.1	5.1	5.4	7.5	13.9	17.1	25.9	9.6	11.5
Distribution of coal use by market sector (%)									
Metallurgical	86	43	41	32	17	17	11	—	—
Electric	0	8	13	11	46	39	58	—	—
Industry	14	35	33	37	20	19	13	—	—
Synthetic Fuels	–	–	–	–	–	4	3	—	—
Residential/Commercial	0	14	13	20	17	21	15	—	—
Total coal use	100%	100%	100%	100%	100%	100%	100%	—	—
Coal consumption/imports (mtce) **Consumption**									
Metallurgical	1.8	2.2	2.2	2.4	2.4	2.8	2.8	1.9	1.9
Steam	0.3	2.9	3.2	5.1	11.5	14.3	23.1	18.3	20.8
Total coal consumption	2.1	5.1	5.4	7.5	13.9	17.1	25.9	9.6	11.5
Imports									
Metallurgical	1.8	2.2	2.2	2.4	2.4	2.8	2.8	1.9	1.9
Steam	0.3	2.9	3.2	5.1	11.5	14.3	23.1	18.3	20.8
Total coal imports	2.1	5.1	5.4	7.5	13.9	17.1	25.9	9.6	11.5
Coal production/exports (mtce) **Production**									
Metallurgical									
Steam									
Total coal production									
Exports									
Metallurgical									
Steam									
Total coal export									

(1) From 1985.

Sweden

II. Coal's Role in Total Energy System	1977	1985		1990		2000		1977-2000 Avg. annual growth–%/yr.	
		A	B	A	B	A	B	A	B
Total Primary Energy Use (mtce)									
Oil, Domestic	0	--	--	--	--	--	--		
Oil, Imported	39	33	38	28	31	22	25	-2.5	-1.9
Gas, Domestic	--	--	--	--	--	--	--		
Gas, Imported	--	--	--	--	--	--	--		
Nuclear	6	14	6	18	1	18	--	4.9	--
Hydro, Solar, Other	20	26	26	29	30	34	34	2.3	2.3
Coal, Domestic	0	0	0	0	0	0	0		
Coal, Imported	2	5	5	7	14	17	26	9.6	11.5
Total energy use	67	78	75	82	76	91	85	1.3	1.0
Coal penetration (%)	3	6	7	9	19	19	30	—	—
Total primary energy (mtce) input to electricity									
Oil and Gas	4	1	4	1	4	1	1	-5.9	-5.9
Hydro, Solar, Other	16	20	20	21	21	22	22	1.4	1.4
Nuclear	6	14	6	18	1	18	--	4.9	--
Coal	0	0	1	1	6	7	15	13.9[1)	19.8[1)
Total energy input	26	35	31	41	32	48	38	2.7	1.7
Coal penetration (%)	0	1	2	2	20	14	39	—	—
Total electric capacity (GWe)									
Oil and Gas	7.9	7.4	7.5	6.0	6.6	3.7	4.7	-3.2	-2.2
Hydro, Solar, Other	13.1	16.2	16.2	16.9	17.1	17.8	18.2	1.3	1.4
Nuclear	3.8	8.4	3.1	9.5	--	9.5	--	4.1	--
Coal	0	0.6	0.9	1.3	6.5	8.2	11.6	19.0[1)	18.6[1)
Total capacity	24.8	32.6	27.7	33.7	30.2	39.2	34.5	2.0	1.0
Coal Penetration (%)	0	2	3	4	22	21	34	—	—
Peak load	5.2	9.7	7.6	7.6	8.9	8.0	8.7	1.9	2.3
Peak reserve margin (%)	21	30	27	23	29	20	25	—	—
Total oil imports (mbd)	0.55	0.52	0.60	0.44	0.49	0.36	0.39	-2.5	-1.9
Total oil consumption (mbd)									
Transportation	0.16	0.18	0.18	0.18	0.18	0.17	0.17	0.3	0.3
Residential/Commercial	0.20	0.17	0.19	0.12	0.13	0.05	0.08	-5.9	-3.9
Industry—Boilers	0.11	0.12	0.13	0.09	0.09	0.09	0.09	-0.9	-0.9
Industry—Other	0.03	0.04	0.04	0.04	0.04	0.04	0.04	1.3	1.3
Electric utilities	0.05	0.01	0.06	0.01	0.05	0.01	0.01	-6.8	-6.8
Total oil consumption	0.55	0.52	0.60	0.44	0.49	0.36	0.39	-2.5	-1.9
World oil price assumed for national coal analysis (1979 U.S. dollars/barrel)	$20*	24	24	28	28	34	34	2.6	2.6
Economic growth assumed for national coal analysis (GNP, billion 1978 dollars)	78	97	97	111	111	142	142	2.6	2.6

* Uses current price of $20/barrel (June 1979 U.S. dollars) as baseline world oil price and as floor price throughout the period.

(1) From 1985.

241

United Kingdom

I. Coal Use, Production, and Trade	1977	1985		1990		2000		1977-2000 Avg. annual growth—%/yr.	
		A	B	A	B	A	B	A	B
Coal use in major markets (mtce)									
Metallurgical	18	17	17	18	21	16	21	–	0.7
Electric	65	68	72	69	73	70	77	0.3	0.7
Industry	9	12	16	14	22	28	47	4.9	7.3
Synthetic Fuels	–	–	–	–	–	4	13	+	+
Residential/Commercial	17	10	10	10	12	15	21	–	0.9
Total coal use	109	107	115	111	128	133	179	0.6	2.2
Distribution of coal use by market sector (%)									
Metallurgical	16	15	15	17	17	12	12	—	—
Electric	59	64	62	62	57	53	43	—	—
Industry	9	11	14	12	17	21	26	—	—
Synthetic Fuels	–	–	–	–	–	3	7	—	—
Residential/Commercial	16	10	9	9	9	11	12	—	—
Total coal use	100%	100%	100%	100%	100%	100%	100%	—	—
Coal consumption/imports (mtce) **Consumption**									
Metallurgical	18	17	17	18	21	16	21	–	0.7
Steam	91	90	98	93	107	117	158	1.0	2.4
Total coal consumption	109	107	115	111	128	133	179	0.7	2.2
Imports									
Metallurgical	1	2	2	2	2	2	2	–	–
Steam	1	–	–	–	–	–	15	–	–
Total coal imports	2	2	2	2	2	2	17	–	–
Coal production/exports (mtce) **Production**									
Metallurgical	16	15	15	16	19	14	19	–	0.7
Steam	92	94	101	96	109	119	143	1.1	1.9
Total coal production	108	109	116	112	128	133	162	0.9	1.7
Exports									
Metallurgical	–	–	–	–	–	–	–	–	–
Steam	1	3	3	3	2	2	–	–	–
Total coal export	1	3	3	3	2	2	–	–	–

United Kingdom

II. Coal's Role in Total Energy System	1977	1985		1990		2000		1977-2000 Avg. annual growth-%/yr.	
		A	B	A	B	A	B	A	B
Total Primary Energy Use (mtce)									
Oil, Domestic	123	103	139	116	144	116	124	--	--
Oil, Imported									
Gas, Domestic	56	77	77	77	77	54	63	--	--
Gas, Imported									
Nuclear	14	27	27	32	32	54	81	6.0	7.9
Hydro, Solar, Other									
Coal, Domestic	109	107	115	111	128	133	179	0.6	2.2
Coal, Imported									
Total energy use	303	313	358	336	380	358	448	0.7	1.7
Coal penetration (%)	36	34	32	33	34	37	40	—	—
Total primary energy (mtce) input to electricity									
Oil and Gas	18	14	19	18	23	14	14	--	--
Hydro, Solar, Other Nuclear	14	27	27	32	32	54	81	6.0	7.9
Coal	65	68	72	69	73	70	77	0.3	0.7
Total energy input	97	109	118	118	127	138	172	1.6	2.5
Coal penetration (%)	67	62	61	58	57	51	45	—	—
Total electric capacity (GWe)									
Oil and Gas Hydro, Solar, Other	17	23	24	22	25	20	20	--	--
Nuclear	5	10	10	14	17	26	40	7.3	9.5
Coal	45	44	44	44	45	44	55	--	0.9
Total capacity	67	77	78	80	87	90	115	1.7	2.4
Coal Penetration (%)	67	57	56	55	52	49	48	—	—
Peak load	50	57	62	62	71	75	93	1.8	2.7
Peak reserve margin (%)	34	35	26	29	23	20	23	—	—
Total oil imports (mbd)**	0.9	(1.1)	(0.7)	(1.0)	(0.6)	0.5	0.6	--	--
Total oil consumption (mbd)									
Transportation									
Residential/Commercial									
Industry—Boilers									
Industry—Other									
Electric utilities									
Total oil consumption **	1.7	1.4	1.8	1.5	1.9	1.5	1.6	--	--
World oil price assumed for national coal analysis (1979 U.S. dollars/barrel)	14½	14½	18	18	23½	25½	35	2.5	3.9
Economic growth assumed for national coal analysis (GNP, billion 1978 dollars)	257	301	326	332	377	405	507	2.0	3.0

** Excluding Oil for Non-Energy Use and Bunkers.

243

United States

I. Coal Use, Production, and Trade	1977	1985		1990		2000		1977-2000 Avg. annual growth—%/yr.	
		A	B	A	B	A	B	A	B
Coal use in major markets (mtce)									
Metallurgical	77	85	85	90	90	100	110	1.1	1.6
Electric	372	500	560	533	748	800	1170	3.4	5.0
Industry	60	70	80	85	112	125	220	3.2	5.8
Synthetic Fuels	0	0	0	5	5	50	200	–	–
Residential/Commercial*	–	–	–	–	–	–	–	–	–
Total coal use	509	655	725	713	955	1075	1700	3.3	5.4
Distribution of coal use by market sector (%)									
Metallurgical	15	13	12	13	10	9	6	—	—
Electric	73	76	77	75	78	74	69	—	—
Industry	11	10	11	12	12	12	13	—	—
Synthetic Fuels	0	0	0	1	1	5	12	—	—
Residential/Commercial	1	1	1	1	1	1	1	—	—
Total coal use	100%	100%	100%	100%	100%	100%	100%	—	—
Coal consumption/imports (mtce) **Consumption**									
Metallurgical	77	85	85	90	90	100	110	1.1	1.6
Steam	432	570	640	623	865	975	1590	3.6	5.8
Total coal consumption	509	655	725	713	955	1075	1700	3.3	5.4
Imports									
Metallurgical	2	5	5	5	5	6	10	–	–
Steam	–	2	2	2	5	5	7		
Total coal imports	2	7	7	7	10	11	17		
Coal production/exports (mtce) **Production**									
Metallurgical	114	130	130	140	145	154	170	1.2	1.6
Steam	443	588	668	651	920	1035	1713	3.8	6.1
Total coal production	557	718	798	791	1065	1189	1883	3.3	5.4
Exports									
Metallurgical	39	50	50	55	60	60	70	1.4	2.0
Steam	11	20	30	30	60	65	130	8.0	11.3
Total coal export	50	70	80	85	120	125	200	3.6	5.8

* Included with industry.

United States

II. Coal's Role in Total Energy System	1977	1985		1990		2000		1977-2000 Avg. annual growth-%/yr.	
		A	B	A	B	A	B	A	B
Total Primary Energy Use (mtce)									
Oil, Domestic	704	772	772	758	758	650	650	--	--
Oil, Imported	650	642	661	657	711	646	708	--	--
Gas, Domestic	655	655	655	625	625	540	540	--	--
Gas, Imported	36	72	72	72	72	72	72	--	--
Nuclear	96	200	220	280	320	300	600	--	--
Hydro, Solar, Other	86	144	144	180	170	327	240	--	--
Coal, Domestic	507	648	718	706	945	1064	1683	--	--
Coal, Imported	2	7	7	7	10	11	17	--	--
Total energy use	2736	3140	3250	3285	3610	3610	4512	1.2	2.2
Coal penetration (%)	19	21	22	22	26	30	38	—	—
Total primary energy (mtce) input to electricity									
Oil and Gas	255	250	250	240	240	175	135	--	--
Hydro, Solar, Other	86	110	110	117	117	150	150	--	--
Nuclear	96	200	220	280	320	300	600	--	--
Coal	372	500	560	533	748	800	1170	--	--
Total energy input	809	1060	1140	1170	1425	1425	2055	2.5	4.1
Coal penetration (%)	46	47	49	46	52	55	57	—	—
Total electric capacity (GWe)									
Oil and Gas	200	214	214	214	214	185	175	--	--
Hydro, Solar, Other	70	82	82	86	86	100	100	--	--
Nuclear	45	100	110	140	160	150	300	--	--
Coal	202	284	305	300	380	465	625		
Total capacity	517	680	711	740	840	900	1200	2.5	3.7
Coal Penetration (%)	39	42	43	41	45	51	51	—	—
Peak load	395	520	570	600	700	760	1020	2.9	4.2
Peak reserve margin (%)	31	31	25	23	20	18	18	—	—
Total oil imports (mbd)	8.5	8.4	8.6	8.6	9.3	8.4	9.2	--	--
Total oil consumption (mbd)									
Transportation	9.7	9.6	9.9	9.7	10.1	9.7	10.8	--	--
Residential/Commercial	3.3	2.8	2.8	2.8	2.8	2.1	2.3	--	--
Industry—Boilers	1.0	1.1	1.1	1.1	1.2	1.0	1.4	--	--
Industry—Other	2.5	2.6	2.6	2.6	2.8	2.4	3.0	--	--
Electric utilities	1.8	2.3	2.3	2.3	2.3	2.0	1.5	--	--
Total oil consumption	18.3	18.4	18.7	18.5	19.2	17.2	19.0	--	--
World oil price assumed for national coal analysis (1979 U.S. dollars/barrel)	$20*	20	25	22.50	30	25	35	--	--
Economic growth assumed for national coal analysis (GNP, billion 1978 dollars)	$1900	2450	2450	2870	2870	3600	3600	2.8	2.8

* Uses current price of $20/barrel (June 1979 U.S. dollars) as baseline world oil price and as floor price throughout the period.

245

Standard Fuel/Energy Equivalences

	mtce/yr	bdoe	10³ bdoe	mtoe/yr	million cfd natural gas	10⁹ m³/yr natural gas	million kWh/yr*	TJ/yr
1 million tce/yr	1	13,121.3	13.121	.646004	76.1048	.7447	8,141	29,307.6
1 bdoe	76.211×10^{-6}	1	10^{-3}	49.2326×10^{-6}	5.8×10^{-3}	56.75×10^{-6}	.620432	2.23356
1 thousand bdoe	76.211×10^{-3}	1×10^{3}	1	49.2326×10^{-3}	5.8	.05675	620.432	2,233.56
1 million toe/yr	1.54798	20,311.8	20.3118	1	117.808	1.1528	12,602.1	45,367.6
1 million cfd natural gas	13.140×10^{-3}	172.414	.172414	8.48837×10^{-3}	1	.009785	106.971	385.096
1 × 10⁹ m³/yr natural gas	1.3428	17,620	17.620	.8675	102.2	1	10,932	39,356
1 million kWh/yr	$.122835 \times 10^{-3}$	1.61178	1.61178×10^{-3}	79.3521×10^{-6}	9.34833×10^{-3}	$.9148 \times 10^{-4}$	1	3.6
1 TJ/yr	34.1208×10^{-6}	.4477	$.4477 \times 10^{-3}$	22.0422×10^{-6}	2.59676×10^{-6}	0.2541×10^{-4}	.277778	1

* Calorific value equivalence of electricity generated in electric power stations by given quantities of fuel would be reduced according to power station thermal efficiency. A multiplier of 35% is a representative figure.

Source: *Coal Pocket Book*, Shell Coal International (1977).

Unit Nomenclature and Standard Abbreviations

Throughout this report the following nomenclature and abbreviations have been used.

bbl	barrel	mbd	million barrels per day
bdoe	barrel(s) per day oil equivalent	mbdoe	million barrels per day oil equivalent
Btu	British thermal unit	mg	microgram
Btu/lb	Brtish thermal unit per pound	MJ	Mega joule = 10^6J
cf	cubic feet	MHD	magnetohydrodynamics
cfd	cubic feet per day	mtce	million tons (metric) coal equivalent
CIF	cost including insurance and freight	mtoe	million tons (metric) oil equivalent
CO	carbon monoxide		
CO_2	carbon dioxide	MWe	megawatt(s) of electric generating capacity = 1000 kWe
CPE's	Centrally Planned Economies		
COM	coal-oil mixture	NGL	natural gas liquids
DWT	deadweight tone	NO_2	nitrogen dioxide
EC	European Community (formerly EEC)	NO_x	nitrogen oxides
FBC	fluidized bed combustion	OBO	ore/bulk/oil combination carrier
FGD	flue gas desulfurization		
FOB	freight on board	OECD	Organization for Economic Cooperation and Development
GDP	Gross Domestic Product		
GNP	Gross National Product		
GRT	gross-registered tons	OPEC	Organization of Petroleum Exporting Countries
GWe	gigawatt(s) of electric generating capacity = 1000 MWe		
		ppm	parts per million
HCV	high calorific value	PPR	Polish People's Republic
IEA	International Energy Agency	PRC	People's Republic of China
IMF	International Monetary Fund	quad	quadrillion (10^{15}) Btu
kcal/kg	kilocalories per kilogram	R & D	research and development
kg	kilogram(s)	SNG	Substitute Natural Gas
km	kilometer(s)	SO_2	sulfur dioxide
km²	square kilometer(s)	SO_x	sulfur oxides
kWe	kilowatt(s) of electric generating capacity	SRC	solvent refined coal
		t	tons (metric)
kWh	kilowatt hour	tce	tons (metric) coal equivalent
lb	pound	tcf	trillion cubic feet
LCV	low calorific value	toe	tons oil equivalent
LDC's	Less Developed Countries	TJ	terajoule = 10^{12} joules
LNG	Liquefied Natural Gas	TSP	total suspended particulates
m²	square meter(s)	tWh	tera watt hours
m³	cubic meter(s)	WEC	World Energy Conference
		yr	year

247